T0210697

Modeling and Data Mining
in Blogosphere

Synthesis Lectures on Data Mining and Knowledge Discovery

Editor
Robert Grossman, *University of Illinois, Chicago*

Modeling and Data Mining in Blogosphere
Nitin Agarwal and Huan Liu
2009

Modeling and Data Mining in Blogosphere

Synthesis Lectures on Data Mining and Knowledge Discovery

Editor
Robert Grossman, *University of Illinois, Chicago*

Modeling and Data Mining in Blogosphere
Nitin Agarwal and Huan Liu
2009

Modeling and Data Mining in Blogosphere Nitin Agarwal and Huan Liu

ISBN: 978-3-031-00770-5   paperback
ISBN: 978-3-031-01898-5   ebook

DOI: 10.1007/978-3-031-01898-5

A Publication in the Springer series
SYNTHESIS LECTURES ON DATA MINING AND KNOWLEDGE DISCOVERY

Lecture #1
Series Editor: Robert Grossman, *University of Illinois, Chicago*

Series ISSN
Synthesis Lectures on Data Mining and Knowledge Discovery
ISSN pending.

Modeling and Data Mining in Blogosphere

Nitin Agarwal
University of Arkansas at Little Rock

Huan Liu
Arizona State University

SYNTHESIS LECTURES ON DATA MINING AND KNOWLEDGE DISCOVERY #1

ABSTRACT

This book offers a comprehensive overview of the various concepts and research issues about blogs or weblogs. It introduces techniques and approaches, tools and applications, and evaluation methodologies with examples and case studies. Blogs allow people to express their thoughts, voice their opinions, and share their experiences and ideas. Blogs also facilitate interactions among individuals creating a network with unique characteristics. Through the interactions individuals experience a sense of community. We elaborate on approaches that extract communities and cluster blogs based on information of the bloggers. Open standards and low barrier to publication in Blogosphere have transformed information consumers to producers, generating an overwhelming amount of ever-increasing knowledge about the members, their environment and symbiosis. We elaborate on approaches that sift through humongous blog data sources to identify influential and trustworthy bloggers leveraging content and network information. Spam blogs or *splogs* is an increasing concern in Blogosphere, which is discussed in detail with the approaches leveraging supervised machine learning algorithms and interaction patterns. We elaborate on data collection procedures, provide resources for blog data repositories, mention various visualization and analysis tools in Blogosphere, and explain conventional and novel evaluation methodologies, to help perform research in the Blogosphere.

The book is supported by additional material, including lecture slides as well as the complete set of figures used in the book, and the reader is encouraged to visit the book website for the latest information:

http://tinyurl.com/mcp-agarwal

KEYWORDS

blogosphere, weblogs, blogs, blog model, power law distribution, scale free networks, degree distribution, clustering coefficient, centrality measures, clustering, community discovery, influence, diffusion, trust, propagation, spam blogs, splogs, data collection, blog crawling, performance evaluation

To my parents, Sushma and Umesh Chand Agarwal . . . –NA

To my parents, wife, and sons . . . –HL

. . . with much love and gratitude for everything.

Contents

Acknowledgments

We thank many colleagues who made substantial contributions in various ways to this book project. The members at the Social Computing Group, Data Mining and Machine Learning Lab at ASU made this project much easier and enjoyable. They include Magdiel Galan, Shamanth Kumar, Sai Moturu, Lei Tang, Xufei Wang, Reza Zafarani, and Zheng Zhao.

We really appreciate Morgan & Claypool and particularly executive editor Diane D. Cerra for helping us throughout this project. The work is part of the projects sponsored by grants from AFOSR and ONR.

Last, and certainly not least, we thank our families, for supporting us through this fun but time-consuming project. We dedicate this book to them, with love.

Nitin Agarwal and Huan Liu
July 2009

CHAPTER 1

Modeling Blogosphere

The advent of participatory Web applications (or Web 2.0 [1]) has created online media that has turned the former mass information consumers to the present information producers [2]. Examples include blogs, wikis, social annotation and tagging, media sharing, and various other such services. A blog site or simply blog (short for web log) is a collection of entries by individuals displayed in reverse chronological order. These entries, known as the blog posts, can typically combine text, images, and links to other blogs, blog posts, and/or to Web pages. Blogging is becoming a popular means for mass Web users to express, communicate, share, collaborate, debate, and reflect. Blogosphere is a virtual universe that contains all blogs. Blogosphere also represents the network of the blogs where each node could be either a blog or a blog post and the edges depict a hyperlink between two nodes in this blog network. Bloggers, the blog writers, loosely form their special interest communities where they share thoughts, express opinions, debate ideas, and offer suggestions interactively. Blogosphere provides a conducive platform to build the *virtual communities* of special interests. It reshapes business models [3], assists viral marketing [4], provides trend analysis and sales prediction [5, 6], aids counter-terrorism efforts [7], and acts as grassroot information sources [8].

Past few years have observed a phenomenal growth in the blogosphere. Technorati (http://technorati.com/blogging/state-of-the-blogosphere/) published a report on the growth of the blogosphere. The report mentioned that the blogosphere is consistently doubling every 5 months for the last 4 years and the size was estimated to be approximately 133 million blogs by December 2008. Furthermore, 2 new blogs or roughly 18.6 new blog posts are added to the blogosphere every second. Given the prominent and continued growth of the blogosphere, it is natural to ask whether it is possible to model the growth of the blogosphere and derive some macro-level statistics that characterizes the blog network. To study the complex network such as blogosphere, researchers can develop blog models and generate data through these models while continuously collecting blog data.

A unique characteristic of blogs is the evolution of its content over time. Unlike traditional web pages that are more or less static (with a few additions, deletions, substitutions, or changes in layout) over time, blogs accumulate content in a prescribed fashion. While the content on web pages is hard to track over time, content is appended to blogs in the form of blog posts and comments which are timestamped in a reverse-chronological order. Another key characteristic of the blogs is the ability of the readers to express their opinions about a blog post interactively by leaving comments. These comments become an integral part of the blog post. Some comments may trigger more comments or a completely new blog post. Moreover, these interactions could be observed within the same blog or across different blogs through references and citations. These timestamped, interaction patterns are highly dynamic, presenting challenges observing the evolution of the blogosphere. Analyzing complex and temporal phenomena often requires the assistance of synthetic data that matches observed characteristics. Generating synthetic data requires developing and fitting a model on the observed data. Such a model can give insights to the interaction patterns in the blogosphere, help in evaluating hypotheses and concepts, study characteristic properties of the network, and predict trends, among others.

1.1 MODELING ESSENTIALS

The blogosphere consists of two main graph structures - a *blog network* and a *post network*. A post network is formed by considering the links between blog posts, and ignoring the blogs to which they belong. In a post network, the nodes represent individual blog posts, and edges represent the links between them. A post network gives a microscopic view of the blogosphere and helps in discerning "high-resolution" details like blog post level interactions, communication patterns in blog post interactions, authoritative blog post based on links, etc. A blog network is formed by collapsing those individual nodes in the post network that belong to a single blog, to a single node. By doing so links between the blog posts that belong to a single blog disappear and links between blog posts of different blogs are agglomerated and weighted accordingly. A blog network gives a macroscopic view of the blogosphere and helps in observing "low-resolution" details like blog level interactions, communication patterns in blog-blog interactions, authoritative blogs based on links, etc. Both post and blog networks are directed graph.

Example 1.1. An example of a post network and a corresponding blog network is displayed in Figure 1.1. A node (orange rectangle) in Figure 1.1(a) represents a blog post and the edges between the nodes are the links. An edge pointing to P_{12} from P_{24} means that blog post P_{24} has an outlink to blog post P_{12}. Each blog post belongs to a blog. For instance, blog posts P_{11}, P_{12}, and P_{13} belong to the blog B_1. Figure 1.1(b) shows the corresponding blog network obtained by collapsing the post network. Each blog represents the node and the links between the blog posts are collapsed to form edges between the blogs. The weight on the edges of a blog graph represents the number of links in the original post network that were collapsed. For instance, the edges from P_{21} to P_{12}, P_{24} to P_{12}, and P_{23} to P_{13} are collapsed to form a single edge from blog B_2 to B_1. Since three links were collapsed to form a single edge, the weight on this edge is 3. Note that the edges between the posts of a blog are not retained in the blog network. For instance, edge from P_{12} to P_{11} does not exist in the blog network.

A third type of network structure in the blogosphere, a *blogger network*, could be derived from the post network. The nodes in the post network could be replaced by the authors of these posts or the bloggers. Nodes that correspond to the same blogger are collapsed. The links between the blog posts form the edges between the corresponding bloggers. A blogger network gives a different interpretation of the blogosphere. It generates a type of social network among bloggers. Note that a blogger network is also a directed graph. However, in this chapter we will restrict our focus to the study of post and blog networks. Most of the models and network characteristics presented here are equally applicable to blogger networks.

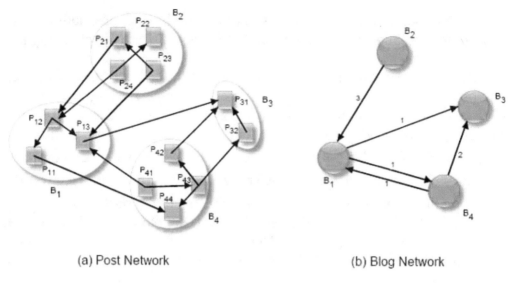

(a) Post Network   (b) Blog Network

Figure 1.1: A post network and the corresponding blog network.

Since the primary difference between a blog network and a post network is the granularity of the information, models developed to simulate either of the two network structures work equally well for the other. Structures of both types of network exhibit the same network characteristics. In the remaining of this chapter, we describe the models and the network characteristics without differentiating the two network structures.

Several characteristic statistics of the blogosphere can be studied and modeled. Observing these statistics helps identifying any unusual behavior in a network and evaluating the proposed models because any model that attempts to simulate the network must be able to reproduce these statistics. Network statistics can be broadly categorized into *stable statistics* (those stabilize over time) or *evolutionary statistics* (those evolve or change dynamically).

Stable Statistics: Stable statistics are time invariant and present a macroscopic information about the network. Examples of stable statistics are degree distribution, clustering coefficient, diameter of the network, average path length of the network etc.

Evolutionary Statistics: Evolutionary statistics are time variant properties of the network and tell us how the network evolves over time. Examples of evolutionary statistics include average lifespan of communities, growth rate of degree of a node etc.

Stable statistics can be further categorized into individual statistics, relational statistics, and global statistics. Each of these categories is explained as follows:

Individual Statistics: Individual statistics provide details about individual nodes of the network, e.g., degree distribution.

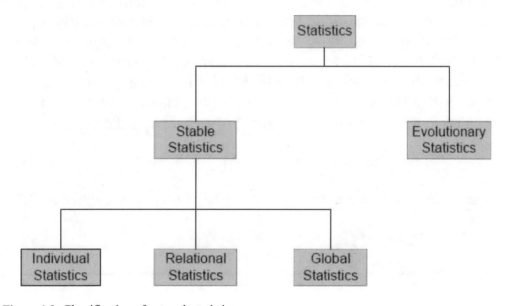

Figure 1.2: Classification of network statistics.

Relational Statistics: Relational statistics provide details about the edges between the nodes or relations, hence the name relational statistics, e.g., clustering coefficient.

Global Statistics: Global statistics provide global information about the network, e.g., diameter of the network, and average path length of the network.

A hierarchical organization of various categories of statistics is depicted in Figure 1.2.

Recently, some additional statistics are proposed to characterize a network such as degree correlation, reciprocity, average degree, total density, community size and density distribution, connectivity, blogger growth and attrition, blogger retention, edge persistence, etc. In this chapter, we discuss the most prominent and widely-used statistics. Specifically, we focus on degree distribution, clustering coefficient, diameter of the network, and average path length of the network.

Degree distribution: Degree of a node in a network refers to the number of edges that are incident upon the node or the number of connections a node has to other nodes. This definition of the degree of a node refers to the case of undirected network. In a directed network, degree of a node is further specified according to edge directions - *indegree* and *outdegree*. Indegree refers to the number of incoming edges of a node, while outdegree refers to the number of outgoing edges from the node.

Formally, for a network $G = (V, E)$ with V being a set of vertices and E a set of edges between the vertices, an edge e_{ij} connects vertices v_i and v_j. For an **undirected network**, degree $d(v_i)$ of a vertex v_i is defined as:

$$d(v_i) = |v_j| \; s.t. \; e_{ij} \in E \wedge e_{ij} = e_{ji} \tag{1.1}$$

Similarly, for a **directed network** indegree ($d_{in}(v_i)$) can be defined as:

$$d_{in}(v_i) = |v_j| \; s.t. \; e_{ji} \in E \tag{1.2}$$

Table 1.1: Indegree and outdegree values for the nodes of the post network in Figure 1.1(a).

Node	Indegree	Outdegree
P_{11}	1	1
P_{12}	2	2
P_{13}	3	1
P_{21}	1	1
P_{22}	1	0
P_{23}	0	2
P_{24}	0	2
P_{31}	3	0
P_{32}	1	1
P_{41}	0	2
P_{42}	1	1
P_{43}	1	3
P_{44}	2	0

and outdegree ($d_{out}(v_i)$) can be defined as:

$$d_{out}(v_i) = |v_j| \ s.t. \ e_{ij} \in E \tag{1.3}$$

Degree distribution is the probability distribution of these degree values over the whole network. Specifically, degree distribution $P(k)$ is defined as the fraction of nodes that have the degree (indegree or outdegree) k. For instance, if there are n nodes in a network ($n = |V|$) and n_k nodes have degree k ($n_k = |v| \ s.t. \ \forall v \in V, \ d(v) = k$), then $P(k) = n_k/n$. In a directed network, we can observe indegree ($P_{in}((k))$ and outdegree ($P_{out}(k)$) distributions separately. Degree distribution gives macroscopic information about the network as compared to the degree values of each node of the network.

Example 1.2. Node P_{13} in Figure 1.1(a) has an indegree of 3 and an outdegree of 1. Table 1.1 lists the indegree and outdegree values of all the nodes in the post network in Figure 1.1(a). Referring to Figure 1.1(a), $P_{in}(1)$ = 6/13 and $P_{out}(1)$ = 5/13. Table 1.2 presents the complete indegree ($P_{in}(k)$)

and outdegree ($P_{out}(k)$) distribution for all possible values of k for Figure 1.1(a).

Clustering coefficient: It quantifies the cliquishness of a vertex with its adjacent vertices or neighbors. Adjacent vertices of a vertex v are those vertices that are directly connected to v. Clustering coefficient of a vertex v takes real values between "0" and "1". A "0" clustering coefficient of v means that there is no edge between any of the adjacent nodes of v. Whereas, a "1" clustering coefficient of v means that all the adjacent vertices of v are directly connected among themselves, hence they form a clique (complete graph). Clustering coefficient of a vertex v is defined as the proportion of edges between the vertices that are adjacent to v divided by the total number of edges that could possibly exist between the adjacent vertices. Clustering coefficient of a network is then defined as the average of the clustering coefficient values of all the vertices in the network.

Table 1.2: Indegree and outdegree distribution of the post network in Figure 1.1(a).

degree: k	$P_{in}(k)$	$P_{out}(k)$
0	3/13	3/13
1	6/13	5/13
2	2/13	4/13
3	2/13	1/13

Let N_v be the set of adjacent vertices of v and k_v be the number of adjacent vertices to node v, where $k_v = |N_v|$. For a directed network, there are $k_v(k_v - 1)$ possible edges that could exist between k_v nodes adjacent to v. This is due to the fact that in a directed network edge e_{ij} is different from e_{ji}. Thus the clustering coefficient C_v for a node v for a **directed network** can be defined as:

$$C_v = \frac{|e_{ij}|}{k_v(k_v - 1)} \quad s.t. \; v_i, v_j \in N_v, e_{ij} \in E \tag{1.4}$$

For an undirected network, there are $k_v(k_v - 1)/2$ possible edges between k_v nodes adjacent to v. This is because in an undirected network edge e_{ij} and e_{ji} are the same. Thus the clustering coefficient C_v for a node v in an **undirected network** can be defined as:

$$C_v = \frac{2|e_{ij}|}{k_v(k_v - 1)} \quad s.t.\ v_i, v_j \in N_v, e_{ij} \in E \tag{1.5}$$

The clustering coefficient for the whole network (C) is defined as the average of clustering coefficients of all the vertices in the network:

$$C = \frac{1}{|V|} \sum_{v \in V} C_v \tag{1.6}$$

Example 1.3. Lets look at the clustering coefficient of the vertex B_4 in the blog network in Figure 1.1(b). B_4 has two adjacent vertices, B_1 and B_3. Hence $k_{B_4} = 2$. Since the blog network is a directed network, the total possible edges that could exist between B_1 and B_3 is given by $k_{B_4}(k_{B_4} - 1) = 2$. From the network we can see that there is only one edge between B_1 and B_3, so the clustering coefficient of B_4, i.e., $C_{B_4} = 1/2 = 0.5$. The clustering coefficient values of all the vertices in the blog network in Figure 1.1(b) is computed and shown in Table 1.3. Then the clustering coefficient for the whole blog network, $C = 3/4 = 0.75$. Note that if the blog network in Figure 1.1(b) is converted to an undirected network by removing the directions of the edges, the clustering coefficient of all the vertices becomes 1 and so is the clustering coefficient of the whole network.

Table 1.3: Clustering Coefficient of the vertices in the blog network in Figure 1.1(b).

v	C_v
B_1	1/2
B_2	1
	1

B_3	
B_4	1/2

Diameter: Diameter of a network is defined as the shortest path between the farthest pair of vertices. This essentially tells us that between any pair of vertices the number of edges to be traversed is always less than or equal to the value computed as the diameter of the network. To compute the diameter we first find the shortest path between all the pairs of vertices and select the path with the maximum length as the diameter of the network. Mathematically, diameter of a network G (directed or undirected) can be defined as:

$$Diameter(G) = \max_{\forall v_i, v_j \in V, \ v_i \neq v_j} p_s(v_i, v_j) \tag{1.7}$$

where $p_s(v_i, v_j)$ denotes the length of the shortest path between vertices v_i and v_j. It is assumed that the network is connected, meaning that one can reach any vertex starting from any other vertex. However, if the network is not connected then one can identify sub-networks that are connected, also known as connected components of the network. Then pick the largest connected component (connected component with largest number of vertices) and compute its diameter.

Example 1.4. Consider the blog network shown in Figure 1.1(b). This network is not completely connected. So we can identify the largest connected component, which in this case consists of the nodes B_1 and B_4. The diameter of this component is 1. However, if we consider the undirected version of this network by omitting the directions of the edges then the diameter of the network would be 2, i.e., the path B_2-B_1-B_4.

Average path length: It is defined as the average number of edges that lie along the shortest paths for all possible pairs of vertices in a network. From the perspective of social networks, average path length tells the average number of individuals one has to communicate through to reach a complete stranger. Shorter average path length is always more favorable.

Mathematically, the average path length $p_{avg}(G)$ of a network G (directed or undirected) can be defined as:

$$p_{avg}(G) = \frac{1}{|V|(|V|-1)} \sum_{\forall v_i, v_j \in V,\ v_i \neq v_j} p_s(v_i, v_j) \tag{1.8}$$

where $p_s(v_i, v_j)$ denotes the length of the shortest path between vertices v_i and v_j. If there is no path between v_i and v_j then $p_s(v_i, v_j) = 0$. Smaller p_{avg} does not necessarily mean that the information will diffuse faster. This is due to the fact that only reachable pair of nodes are considered while computing p_{avg}. A network that is largely disconnected could still have a short average path length. Therefore, only connectedness and average path length together can indicate the speed of information diffusion (later in Chapter 3).

Example 1.5. Consider the blog network shown in Figure 1.1(b). We compute the shortest paths between all possible pairs of vertices as shown below. The average path length p_{avg} for this network is $9/12 = 0.75$.

$$
\begin{array}{lll}
B_1 \rightarrow B_2 = 0, & B_1 \rightarrow B_3 = 1, & B_1 \rightarrow B_4 = 1 \\
B_2 \rightarrow B_1 = 1, & B_2 \rightarrow B_3 = 2, & B_2 \rightarrow B_4 = 2 \\
B_3 \rightarrow B_1 = 0, & B_3 \rightarrow B_2 = 0, & B_3 \rightarrow B_4 = 0 \\
B_4 \rightarrow B_1 = 1, & B_4 \rightarrow B_2 = 0, & B_4 \rightarrow B_3 = 1
\end{array}
$$

1.2 PREFERENTIAL ATTACHMENT BLOG MODELS

Now we look at the *preferential attachment* model that closely simulates the statistics observed in real blogosphere dataset. According to preferential attachment model, a node of higher degree has higher chance to get new edges. Mathematically, the probability $P(e_{ij})$ of an edge between vertices v_i and v_j is proportional to the number of edges incident upon v_i, as shown below:

$$P(e_{ij}) \propto \frac{d(v_i)}{|V|} \tag{1.9}$$

An undirected network following the preferential attachment model can be generated as follows: a new vertex (v_j) to be added to the network creates an edge with a node v_i with the probability proportional to the number of edges incident upon v_i, as shown in Eq 1.9.

In a directed network, the probability ($P(e_{i \leftarrow j})$) with which a node v_j will create an outlink to a node v_i is proportional to the number of links already pointing to v_i or the number of inlinks of v_i. Mathematically,

$$P(e_{i \leftarrow j}) \propto \frac{d_{in}(v_i)}{|V|} \qquad (1.10)$$

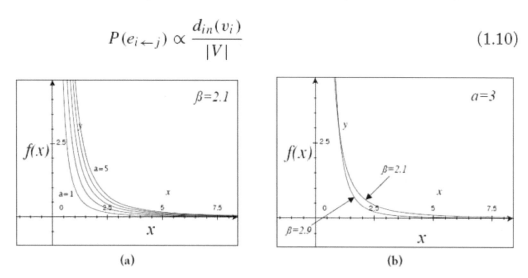

(a)   (b)

Figure 1.3: Basic form of the Power Law function with (a) variable a = {1, 2, 3, 4, 5}, and (b) variable β = {2.1, 2.9}.

Similarly, the probability ($P(e_{i \rightarrow j})$) with which a node v_j will acquire an inlink from a node v_i is proportional to the number of links coming out of v_i or the number of outlinks of v_i. Mathematically,

$$P(e_{i \rightarrow j}) \propto \frac{d_{out}(v_i)}{|V|} \qquad (1.11)$$

A directed network following the preferential attachment model can be generated as follows: an edge from a new vertex v_j to the existing vertex v_i is created with the probability proportional to the number of inlinks of v_i, as shown in Eq 1.10. Here the node v_j acts as a *source vertex* and this models the outlink generation for a new vertex to be added to the network. Similarly, inlink generation for a new vertex v_j can be modeled using Eq 1.11. Any existing node v_i points to a new node v_j with a probability proportional to the number of outlinks of v_i. v_j acts as a *sink vertex*.

Example 1.6. Consider the post network shown in Figure 1.1(a). A new vertex in the network would most likely create a outlink to vertices P_{13} or P_{31} since they have the highest indegree (3). Similarly, the new vertex would most likely acquire an inlink from vertex P_{43} since P_{43} has the highest outdegree (3).

According to the preferential attachment model, vertices with larger degree tend to attract more edges. This creates a *"rich get richer"* phenomenon. This phenomenon has also been widely studied in complex networks, scientific productivity, journal use, "cumulative advantage" in citation learning, wealth distribution domain, etc. Preferential attachment model generates a power law distribution,

$$f(x) = ax^{-\beta} \tag{1.12}$$

where a and β are constants, x is the characteristic that is simulated using the preferential attachment model and f models the distribution of x. If x is the degree of the network, then $f(x)$ models the degree distribution. Figure 1.3(a) shows the basic form of power law with parameters $a = \{1, 2, 3, 4, 5\}$ and $\beta = 2.1$. This shows that the lower the value of a the steeper the fall in the power law distribution. Figure 1.3(b) shows the basic form of power law with parameters $a = \{3\}$ and $\beta = \{2.1, 2.9\}$. This shows that the higher the value of β the steeper the fall in the power law distribution. Power law generates a heavy tailed distribution. β in Eq 1.12 is also called the scaling exponent. This means that $f(cx) \propto f(x)$, where c is a constant. Thus if the function's argument is rescaled the basic form still remains intact except changing the constant of proportionality, as shown below after substituting x with cx in Eq 1.12,

$$f(cx) = a(cx)^{-\beta} = \frac{a}{c^{\beta}}x^{-\beta} = a'x^{-\beta} \tag{1.13}$$

The above derivation shows that scaling the argument of the function from x to cx does not affect the basic form of the power law function. Notice that the function has a linear relationship with slope β. This linear form is extremely helpful in developing a power law model for degree distribution of the blogosphere. Figure 1.4 shows the outdegree distribution of a set of blogs collected from Blogcatalog.com. The x-axis shows the log of the outdegree value and y-axis shows the log of the frequency of the outdegree values. As shown in Figure 1.4 the outdegree distribution of the blogs follow a power law distribution with a linear fit, the slope of which is equal to the scaling exponent $\beta = 1.1693$. We also study the indegree (follower) and outdegree (friend) distribution of a popular microblogging site, Twitter. Both indegree and outdegree distributions follow power law as depicted in Figure 1.5(a) and (b), respectively. Based on the linear fit on the log-log scale of the power law distributions, we can estimate the scaling exponent for indegree and outdegree distributions as 1.1517 and 1.3836, respectively.

An interesting property of power law distribution is its scale invariance or the ability to generate a scale free network. Scale invariance also refers to the self-similarity property where the sub-network is similar to the whole network in terms of network statistics. One way to identify the scale invariance property is to extract a subnetwork by performing a E-R (Erdos-Renyi) random walk on the whole network and then observe the degree distribution of the subnetwork. Blogosphere has been found to exhibit the scale invariant property as also simulated by power law models.

Other characteristics of the blogosphere such as high clustering coefficient, short average path length, and small network diameter can be simulated using the preferential attachment model. High clustering coefficient indicates that often bloggers communicate within a small focused group as reflected by their link interactions, hence forming a sub-community structure. Preferential attachment model generates a network with high clustering coefficient. Blogosphere is typically characterized by short average path length, which indicates that, on average, one can reach a blog from another blog by following the links in small number of hops. Preferential attachment model generates networks with short average path length. Blogosphere is also characterized by small network diameter, i.e., even the furthest pair of blogs are not very far. The network thus generated through a preferential attachment model exhibits small network diameter. These characteristics are explained by the small-world phenomena and such networks are also known as small-world networks. Here most vertices are not neighbors of one another but most vertices can be reached from every other vertex by a small number of steps or hops. This is because the shortest path between vertices passes through highly connected vertices also the *hubs*. These hubs ensure that long range vertices could be reached within small hops.

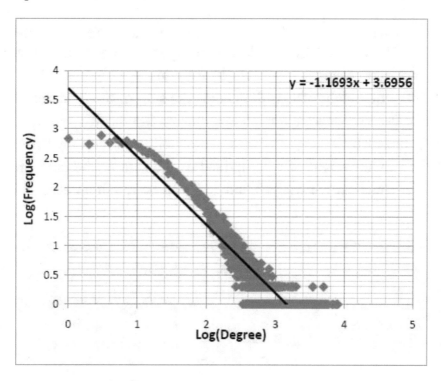

Figure 1.4: Outdegree distribution of Blogcatalog [9].

Preferential Attachment Models - Hybrid variant: A preferential attachment model always favors the vertex with higher degree, hence making the rich even richer. The hybrid variant tries to give a lucky poor vertex a chance by randomly selecting a vertex to link to. The probability of forming an edge between a new vertex v_j and existing vertex v_i is proportional to a linear sum of the degree of vertex v_i and a uniform random probability ϵ [10], as shown below:

$$P(e_{ij}) \propto \gamma \frac{d(v_i)}{|V|} + (1 - \gamma)\epsilon \qquad (1.14)$$

Here $\gamma \leq 1$ tunes the contribution of the preferential attachment model and the uniform random model. This also attempts to generate a reducible network by preventing formation of isolated subnetworks

Preferential Attachment Models - α variant: The α variant, also known as α-scheme, is another modification of the preferential attachment model [11]. In a directed network, according to α-scheme, a new vertex v_j acting as a source vertex connects to a vertex v_i proportional to the in-weight of v_i. In-weight of v_i is defined as the sum of the indegree of v_i and α, where α is a user-specified model parameter. Similarly, v_j acting as a sink vertex is linked by a vertex v_i with a probability proportional to the out-weight of v_i. Out-weight of v_i is defined as the sum of the outdegree of v_i and α. Mathematically, Eqs 1.10 and 1.11 can be rewritten according to α-scheme as:

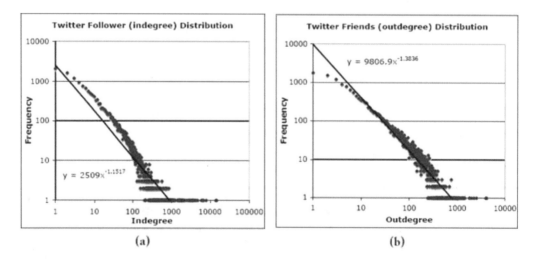

Figure 1.5: Degree distribution of Twitter (a) Indegree and (b) Outdegree.

$$P(e_{i \leftarrow j}) \propto \frac{d_{in}(v_i) + \alpha}{|V|} \tag{1.15}$$

$$P(e_{i \rightarrow j}) \propto \frac{d_{out}(v_i) + \alpha}{|V|} \tag{1.16}$$

In an undirected network, α-scheme is similarly defined as shown below:

$$P(e_{ij}) \propto \frac{d(v_i) + \alpha}{|V|} \tag{1.17}$$

Note that the α-preferential attachment model can be viewed as adding α edges in the network. This could be done to reduce the sparsity in the network. $\alpha = 1$ is a special case and often referred as 1-preferential attachment model.

1.2.1 LOG-NORMAL DISTRIBUTION MODELS

Recently, more sophisticated models like Log-Normal distribution models and Double Pareto Log-Normal distribution models have been studied to simulate the social networks such as the ones emerging from discussion forums and mobile call graphs, respectively. However, their application to the blogosphere still remains an area of research. Here we briefly introduce these models.

The log-normal distribution is a single tailed probability distribution of a random variable whose logarithm is normally distributed, hence the name log-normal. The probability density function of the log-normal distribution has the following form:

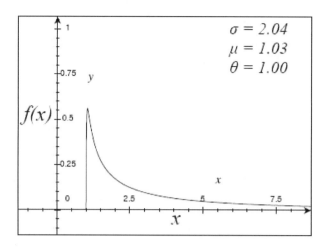

Figure 1.6: A typical form of the Log-Normal distribution function with $\sigma = 2.04$, $\mu = 1.03$, and $\theta = 1.00$.

$$f(x) = \frac{1}{(x - \theta)\sigma\sqrt{2\pi}} \exp\frac{-(\ln(x - \theta) - \mu)^2}{2\sigma^2} \qquad (1.18)$$

where μ and σ are the mean and standard deviation of the distribution, respectively; and θ simply defines the shift in the distribution, which in case of degree distribution represents a lower bound of the degree values. To determine the maximum likelihood estimators of the parameters μ and σ of the log-normal distribution, the same procedure can be used as the normal distribution. The estimated values of these parameters are given by:

$$\hat{\mu} = \frac{\Sigma_k \ln x_k}{n}, \quad \hat{\sigma}^2 = \frac{\sigma_k (\ln x_k - \hat{\mu})^2}{n} \qquad (1.19)$$

A basic form of the log-normal distribution is shown in Figure 1.6 with the values of the parameters $\sigma = 2.04$, $\mu = 1.03$, and $\theta = 1.00$. Gomez et al. [12] used log-normal distribution to model the degree distribution on one of the popular discussion forums, Slashdot.org. They concluded that log-normal is a better fit in terms of Kolmogrov-Smirnov statistical tests for Slashdot's degree distribution as compared to the power law.

In Figure 1.7(a) and (b), we compared the log-log plots of power law and log-normal distributions. Figure 1.7(a) represents the log-log plot of the power law shown in Figure 1.3(b) with $a = 3$ and $\beta = 2.1$. Similarly, Figure

1.7(b) depicts the log-log plot of the log-normal distribution shown in Figure 1.6. The linear relation of log-log plot for power law in Figure 1.7(a) is quite different from the parabolic nature of the log-log plot of the log-normal distribution as shown in Figure 1.7(b).

Double pareto log-normal (DPLN) distribution is a sophisticated version of the log-normal distribution functions and can be considered as a mixture of log-normal distributions. The parametric form of DPLN distribution is much more complex than log-normals. As the models get more and more complex they might run into the risk of overfitting the data. The distinct features of DPLN distribution are two linear sub-plots in the log-log scale and a hyperbolic middle section. DPLN distribution has been used to model the properties of the mobile call graphs such as: number of calls, distinct call partners, and total talk time and have been proved to outperform power law and log-normal distributions. However, the applicability of DPLN to the blogosphere needs further exploration.

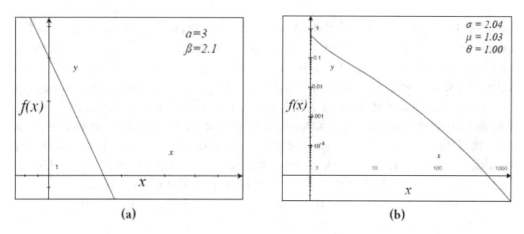

Figure 1.7: Log-log plots of the (a) power law distribution and (b) log-normal distribution.

CHAPTER 2

Blog Clustering and Community Discovery

Chapter 1 discussed the phenomenal growth of the blogosphere. According to the yearly reports published by Technorati, Blogosphere has a growth rate of 100% for every 5 months consistently for the last 4 years. With overwhelming amount of data, it is really difficult to keep track of the state of the blogosphere. Tools are needed to help users organize and make sense of the blogs and blog posts in the blogosphere. Clusters of similar blogs help identify the topics to gain a high-level understanding of the blogosphere. The reader can focus on the cluster he is interested in. The automatic organization of the blogs can also benefit search engines. For example, organized content in the blogosphere can improve indexing mechanisms of the search engines. Clustering can also help improve the delivery of the search results by organizing similar results together. Instead of laying out all the results as a list, similar results can be clustered together and the most dominant topic among the cluster members can be assigned as the label for the cluster. This greatly enhances the navigability of search results. This concept has been studied in webpages, e.g., Clusty (http://www.clusty.com), but remains to be explored in the blogosphere.

Clustering can be used to suggest related content to other cluster members, hence improving recommendation engines. A blogger would be interested to find other bloggers who share similar interests. Clustering of blogs would put all such bloggers in one group who could share, collaborate, and discuss their views, opinions, or perspectives over products, policies, events, or government actions. These blog clusters can also be treated as communities or focussed groups. However, this approach of community discovery is purely based on the content similarity. Nevertheless, bloggers can explicitly specify connections to other fellow bloggers via hyperlinks that connect their blogs or blog posts (as also shown in Figure 1.1), or specify their friends using social network capabilities available at some blog sites. This creates a blog or a blogger network with nodes depicting the blogs or the bloggers and edges indicating the connections among them. One way to find communities is to use the principle of *homophily*, which means that two people tend to communicate more often if they share similar views. Using this phenomenon, those bloggers who have more edges within themselves can be considered as a community. Identifying a set of bloggers that communicate more often among them implies that they share similar views, opinions, or interests; hence they form a community. Obviously, this approach to community discovery is purely based on the network information. It can be extended to a tool for advertisers and marketeers, for whom a global view of likes, dislikes, and interests of groups of bloggers matter.

The Web and the blogosphere are often compared, and the existing approaches in webpage clustering or web community discovery are explored for their use in similar tasks in the blogosphere. But there is a key difference between the blogosphere and the Web in terms of the lifetime of their contents (pages and links) posted on both media. Blogs are mainly used as a tool for timely communication. So the entries in the blogs are often short-lived. Most of them become obsolete and are never referred later. Thus the links in the blogs have significant temporal locality. However, in the Web, new pages may refer to very old pages (e.g., an authoritative webpage) creating a longer lifetime of content in the Web. We may also aggregate the links in the blogs over time, but miss the key temporal dynamics observed in the blogosphere. Bloggers' interests shift over time. For instance, a blogger is initially interested in "politics" so he interacts more often with fellow bloggers who are also interested in "politics", forming a community. Later his (her) interest shifts to "economics" so he/she shifts the interactions with the bloggers who are also interested in "economics". Now if all the interactions are aggregated over time, we would lose this temporal dynamics in the interaction patterns and community evolutions. Based on this key difference between the Web and the blogosphere, conventional webpage clustering or community discovery algorithms do not work well in the blogosphere domain. In this chapter, we will explore some of the techniques that focus on the dynamics of interactions or the evolution of blog communities.

Since blogs are a collection of blog posts, it is often a moot point whether to cluster blog posts and extract blog clusters. However, this approach suffers from the following key challenges:

1. **Content perspective:** Often bloggers discuss various things in different blog posts ranging from daily routines to their opinions on a matter. It is extremely hard to identify the main theme of the blog through a single blog post. However, considering a collection of blog posts from the same blog lets us glean the blog's main theme or the blogger's main interests.

2. **Network perspective:** Due to the casual nature of the blogosphere not many bloggers cite the original source that inspired them to write their blog post, or if they reused content from some other blog or news source. This leads to an extremely sparse blog post network. However, considering a collection of blog posts from the same blog forms a critical mass of such links and the network becomes less sparse.

As mentioned earlier, blogs are associated with content and network information, both of which are extremely useful for clustering and community discovery. Although both content and network information can be used to identify blog clusters or blog communities, it is more prevalent to leverage content information to identify blog clusters and network information to identify blog communities. We illustrate this observation in Figure 2.1. However, this does not stop content information to be used in conjunction with the network information to identify blog communities or network information in conjunction with content information to identify blog clusters. In this chapter, we will first discuss network based and content based approaches to perform clustering and community discovery, and then we will discuss hybrid approaches leveraging both content as well as network information.

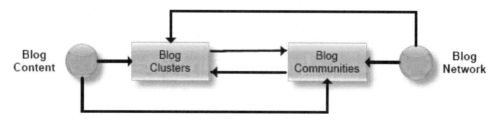

Figure 2.1: Difference and similarities between blog clustering and blog community discovery tasks.

2.1 GRAPH BASED APPROACH

Given a blog network it is natural to represent it as a graph with blogs as nodes and hyperlinks between the blogs as edges. An illustration of the blog network is shown in Figure 1.1(b). The nodes of this graph are the blogs and the edges are the hyperlinks. As mentioned earlier, blogs interact more often with other blogs that share similar opinions or ideas. This leads to more hyperlinks or edges within few blogs. A graph based approach tries to identify such sets of blogs with more links within. These sets of blogs are referred as communities. However, given the sparse nature of hyperlink graph in the blogosphere, often it is challenging to identify communities through the hyperlink graph. Later in this section, we will discuss ways to infer blog network.

Let us assume that the blog graph is represented by $G = (V, E)$, where V is the set of vertices or the blogs in the blog network and E is the set of edges between the blogs. These edges denote the hyperlinks between the blogs. An adjacency matrix W can be used to represent the edges. Note that W can be either a binary matrix denoting the presence or absence of an edge between two blogs or W could be a weighted adjacency matrix denoting the number of links between two blogs as shown in Figure 1.1(b). The weight of the edge between blog v_i and v_j is given by $W(i, j)$. In case of a weighted adjacency matrix, the edge weights are normalized between 0 and 1.

Given the adjacency matrix W, we compute the graph Laplacian (L) of W as the basic step for spectral clustering. The graph Laplacian is computed as:

$$L = D - W \qquad (2.1)$$

where D is the diagonal matrix consisting of entries $d_1, d_2, \ldots, d_n$. Each d_i denotes the degree of the node v_i and is computed as the row-sum of i-th row of W,

$$d_i = \sum_{j=1}^{n} w_{ij} \qquad (2.2)$$

Here n is the number of blogs in the blog graph. Note that this definition of degree is very similar to the definition of degree for undirected graph presented in Chapter 1 in Eq 1.1.

Next, we compute the eigenvalue decomposition of the matrix L and pick the first k eigenvectors $e_1, e_2, \ldots, e_k$ such that the corresponding eigenvalues are in increasing order. Each eigenvector is $n \times 1$. The k eigenvectors are juxtaposed to construct a matrix of $n \times k$ dimension. Each row of this matrix represents a blog in k-dimensional space. We can run k-means clustering algorithm to compute k clusters. This procedure of graph partitioning through spectral clustering is shown in Algorithm 1. Note that L is not normalized in this algorithm. However, L can be normalized to norm 1 before computing the clusters [13, 14].

Input : Adjacency matrix: W,
       Number of clusters: k
Output: k clusters of n nodes/blogs in the blog graph

1 Compute the diagonal matrix, D;
2 Compute the graph laplacian, $L = D - W$;
3 Compute the first k eigenvectors, $e_1, e_2, \ldots, e_k$ of L;
4 Juxtapose these eigenvectors to construct a $n \times k$ matrix;
5 Compute k clusters using k-means algorithm on this matrix;

Algorithm 1: Algorithm for spectral clustering.

Example 2.1. Lets consider an example to illustrate the functioning of spectral clustering algorithm presented in Algorithm 1. Consider a blog network presented in Figure 2.2(a). For simplicity, we consider an unweighted blog network. Each edge in the network assumes an equal weight of 1. Based on this blog network, we compute a binary adjacency matrix, W, as shown in Figure 2.2(b). Now using Eq. 2.2, we compute the matrix D, as shown in Figure 2.2(c). Graph Laplacian L of matrix W is computed using Eq. 2.1, as shown in Figure 2.2(d). Eigenvalue decomposition of L is performed and first two eigenvectors are shown in Figure 2.2(e). Each blog in the blog network is represented in terms of the first two eigenvectors and visualized in Figure 2.2(f). Based on this visualization, it can be observed that blogs $\{B_1, B_2, B_3, B_4\}$ and $\{B_5, B_6, B_7,$

B_8, B_9} form two separate communities. Similar communities were obtained using k-means algorithm.

Spectral clustering on graphs generate clusters such that the edges between nodes belonging to different clusters have low weight (or, more dissimilar) and edges between nodes belonging to the same cluster have high weight (or, more similar). This is also the optimization criterion for graph-cut based clustering approaches. RatioCut [15] and Ncut [14] are two variations for graph-cut approach that circumvents the problem of single-member clusters, which the conventional graph-cut approach suffers from. They circumvent the problem by considering the size of the clusters (in terms of number of cluster members-RatioCut, or total weight of edges in the cluster-Ncut) in the objective function. It can be easily shown that spectral clustering based approach can be derived as an approximation to graph-cut approaches such as RatioCut and Ncut [16]. The groups of nodes that are obtained using any of these techniques correspond to the communities. the derived communities of blogs could also be treated as blog clusters.

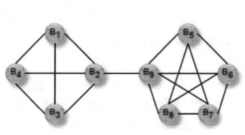

(a) Blog Network

	B_1	B_2	B_3	B_4	B_5	B_6	B_7	B_8	B_9
B_1	0	1	1	1	0	0	0	0	0
B_2	1	0	1	1	0	0	0	0	1
B_3	1	1	0	1	0	0	0	0	0
B_4	1	1	1	0	0	0	0	0	0
B_5	0	0	0	0	0	1	1	1	1
B_6	0	0	0	0	1	0	1	1	1
B_7	0	0	0	0	1	1	0	1	1
B_8	0	0	0	0	1	1	1	0	1
B_9	0	1	0	0	1	1	1	1	0

(b) Matrix: W

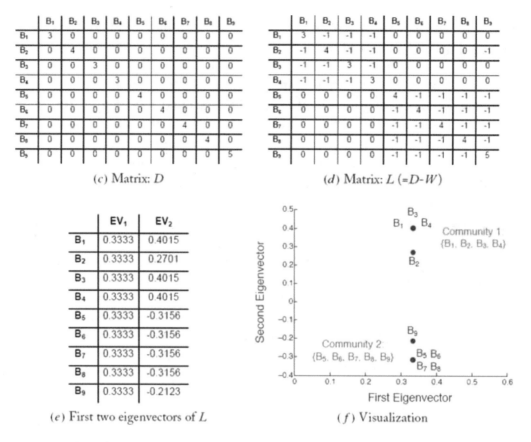

(c) Matrix: D (d) Matrix: $L (=D-W)$

(e) First two eigenvectors of L (f) Visualization

Figure 2.2: A running example of spectral clustering algorithm on a toy dataset.

As mentioned earlier, a blog graph could be extremely sparse due to the casual nature of bloggers. In such a scenario edges between blogs could be derived using content similarity. Given the similarity or distance matrix Q between blogs, there are several ways to construct W.

ϵ-**neighborhood graph:** All those nodes whose pairwise distances are smaller than ϵ are connected in this scheme. Specifically, if Q is the pairwise distance matrix between blogs, then two blogs v_i and v_j are connected by an edge if $Q(i, j) < \epsilon$. The weight of the edge is same as the distance from the matrix Q. Note that ϵ-neighborhood graph can be easily generated for similarity matrix as well. If Q_s is the similarity matrix between blogs then, we make an edge between two blogs v_i and v_j if $Q_s (i, j) > \epsilon$.

k-nearest neighbor graph: k blogs that are nearest to v_i are connected to v_i. Specifically, if a blog v_j is among the k nearest neighbors of v_i, then v_j is connected to v_i. This means that there will be an edge between v_i and v_j with the weight $W(i, j)$ equal to the similarity between v_i and v_j. Note that this is an asymmetric graph because the "nearest" relationship is asymmetric. v_j might be nearest to v_i but v_i might not be nearest to v_j. There could be a node k, which is nearest to j. This leads to a directed graph. One way to make this graph undirected is to drop the directions and treat each edge as bi-directional. Another way to make it undirected is to consider only those edges that connect mutually nearest neighbor nodes/blogs. This means that two blogs, v_i and v_j, will be connected with an edge if both of them are among the k-nearest neighbors of each other.

Fully connected graph: All the blogs are connected with each other with the edge weights equal to the similarity between the blogs. Specifically, blogs i and j are connected with an edge of weight $W(i, j)$ = $Q(i, j)$.

Blogosphere is highly dynamic in terms of the content and interests of the bloggers. Aggregating the interactions of the bloggers over time would miss the dynamics in the interaction patterns and community evolution. Authors in [17] explore the temporal aspects of spectral clustering. The objective function encodes a regularization parameter (also known as the temporal cost) that enforces temporal smoothness. Authors propose two cost functions. One cost function aims to preserve cluster quality by applying the current partition to historical data and the resulting cluster quality determines the temporal cost. The second cost function aims to preserve cluster membership by comparing the current partition with the historical partition and the resulting differences determines the temporal cost. Associating the temporal cost in the objective function ensures that the spectral clustering approach is stable and less sensitive to short-term variations or noise, while at the same time being adaptive to long-term interest drift or community evolution.

The authors in [18] present a novel incremental spectral clustering framework for studying community evolution. Whenever there is a change in the matrix, *W, L* and *D* are incrementally updated. Similarly, the eigenvectors and eigenvalues are incrementally updated. First, *k* smallest updated eigenvectors are picked and *k* clusters are computed using the *k*-means algorithm. Another approach to handle the dynamics of blog communication while extracting communities involves a tensor based approach [19]. The blog network is represented as a link matrix W_i at timestamp *i*. Various such link matrices at different timestamps are stacked together to construct a tensor. The nonnegative matrix factorization of the data tensor results in the community discovery in the dynamic blog network.

2.2 CONTENT BASED APPROACH

Blogosphere has an overwhelming amount of content besides the network or link information. Bloggers express their opinions, share their views, discuss various products, policies, etc. This creates a humongous pool of intelligently crafted content, presenting ample opportunities to explore content based clustering approaches. Content based clustering on Blogosphere has also been studied from text or webpage clustering perspective. However, the inherent differences between webpages and blogs demand novel approaches geared specifically towards blogs. suffer from the sparse link structure prevalent in the blogosphere.

There is a constant stream of data in blogs that creates a highly dynamic environment. Each blog post is timestamped and appears in reverse-chronological order. Blog posts that are significantly distant in time could be very different in terms of content or the central theme. Precisely, this is because blogs are highly motivated by the events happening at that time. This is very different from the webpages. Most webpages are static in terms of content and act as a source of information. They could be updated at times but the theme remains pretty much the same. This difference makes huge impact in clustering approaches. Because of such dynamics, blog posts from a blog cannot be aggregated over time and treated as a big text document or a single webpage for clustering. Doing so will risk the loss of temporal information in the content of blogs and eventually fail to discover community evolution.

Another significant challenge in clustering blogs has to do with the colloquial usage of the language. Given the casual nature of the blogosphere, it is quite natural to observe a lot of misspellings, acronyms, and slangs in the blog content. This presents a huge information quality barrier. Under normal circumstances, such words would be considered as noise and jettisoned. However, in blogs, these words could be as informative as any other words. So one has to be conservative and retain such words. Nevertheless, content based blog clustering suffers from typical text clustering problems, i.e., high-dimensionality and sparsity. We need to pre-process the data in order to remove the noise yet retain informative but seemingly noisy words. Next, we present some pre-processing steps that can specifically help improve information quality:

- Resolve acronyms and casual usage of language using the Urban Dictionary (http://www.urbandictionary.com/). It is a slang dictionary edited by the people and contains 3,873,472 definitions (by the time of writing this book) of the slangs since 1999.

- Remove all the non-content bearing *stopwords* like "a", "an", "the", etc. A list of stopwords can be found at
  http://www.dcs.gla.ac.uk/idom/ir_resources
  /linguistic_utils/stop_words

- Stem the words to retain the roots and discard common endings. This can be achieved by using porter stemmer tool (http://tartarus.org/$\sim$martin/PorterStemmer/).

- Rank the words based on their *tfidf* scores. Pick top-*m* words where *m* could be determined experimentally. This would essentially select distinctive words, i.e., those words that are neither too frequent in all the blog posts nor too infrequent. Note that a *tfidf* score consists of two parts: *tf* and *idf*,

$$tf = \frac{d_i}{\sum_{k=1}^{m} d_k} \tag{2.3}$$

where d_i is the frequency of term i in blog post d and the denominator is the number of occurrences of all terms.

$$idf = \log \frac{d}{t_i} \tag{2.4}$$

where d is the total number of blog posts and t_i is the number of blog posts that contain term i.

$$tfidf = tf \times idf \tag{2.5}$$

A *tfidf* score is normalized between "0" and "1".

- Represent each blog post by a document-term vector also known as the vector-space model. Each row is a blog post and each column in this vector represents a term and the value is the *tfidf* score for that particular blog post.

Such a document-term vector representation is extremely sparse, and it gets even sparser since the individual blog posts are considered independent entities. Sparsity and high-dimensionality leads into various problems like computing the distance between different vectors efficiently. So a common solution to this problem is to transform the document-term vector representation to a concept space vector model using Latent Semantic Analysis (LSA).

Latent semantic analysis computes the singular value decomposition (SVD) of the document-term vector. SVD decomposes the document-term matrix into U, Σ, V matrices as follows:

$$A = U \Sigma V^T \tag{2.6}$$

where A^T is the document-term matrix and V is the document-concept matrix. Matrix V is the reduced concept space representation. This solves the problems that arise due to high-dimensionality and sparsity.

Example 2.2. To give a flavor of how latent semantic analysis creates concept vectors, we consider a hypothetical scenario where blog posts primarily discuss about *stock, investing, cars, trucks, books,* and *harry potter*. Performing LSA would result in linear combination of certain terms mentioned here into a single dimension reflecting something known as a concept. For example, the resulting dimensions might look like,
{(*stock*), (*investing*), (*cars*), (*trucks*), (*books*), (*harry potter*)} → {(1.1476 × *stock* + 0.3498 × *investing*), (1.3452 × *cars* + 0.2828 × *trucks*), (1.2974 × *books* + 0.7391 × *harry potter*)}.

In this hypothetical scenario, (1.1476 × *stock* + 0.3498 × *investing*) component could be interpreted as "stock market", (1.3452 × *car* + 0.2828 × *truck*) component could be interpreted as "vehicle", and (1.2974 × *books* + 0.7391 × *harry potter*) component could be interpreted as "story books". Note that these weights are hypothetical.

Each row in V represents a blog post in the reduced concept space transformation. To compute the distance/similarity between any two pair of blogs, their respective vectors are considered and either cosine similarity is computed or euclidean distance is computed. This similarity or distance matrix is fed into various clustering algorithms like k-means or hierarchical to obtain clusters of blog posts. Most recent blog posts are considered for computing blog post clusters. Clusters of blogs are derived by observing the cluster membership of the blog posts belonging to that blog. This ensures that the dynamics and community evolution are preserved. It has been shown that cosine similarity gives better results for clustering as compared to euclidean distance when dealing with sparse vectors. This version of k-means clustering that uses cosine similarity instead of Euclidean distance to compute the similarity between two data instances, (blog posts in this case) is known as "spherical k-means". have applied spherical k-means to blog clustering problem. Another work by [20] segments the

Blog posts can be segmented into various entities such as title, body or blog post content, and comments [20]. These entities are represented in separate document-term vectors. A single document-term vector is constructed from different entities using a weighted linear combination. Placing higher weights for comments gives best clustering results. Various clustering algorithms have been compared in context of blog clustering and hierarchical clustering has been reported as the best algorithm for content based clustering of blogs [21].

Another way to tap the dynamism of the blogosphere for clustering the blogs is by leveraging the *collective wisdom*. Collective wisdom, a concept studied by psychologists, sociologists and in philosophy, refers to the shared knowledge arrived at by individuals and groups that is used to solve problems. Here the collective wisdom of bloggers is used in clustering the blogs. Bloggers provide annotations or tags for their blogs, assuming that they annotate their blogs with closest possible or most relevant tags. In this work, authors leverage the bloggers' wisdom collectively to develop a knowledge source, which is also represented as a *label relation graph* [22]. The nodes in this graph are the tags or the labels used by the bloggers to annotate their blogs. The edges between two tags (t_i and t_j) denote if they were used simultaneously, and the weights on the edges of the label relation graph denote the number of bloggers that labeled their blogs with both the tags t_i and t_j. This helps in discovering many interesting, complex, and latent relations between tags. A snippet of the label relation graph is shown in Figure 2.3. Note that the relations in the label relation graph are highly dynamic since they change with the change in usage of these tags by the bloggers. This collective wisdom based knowledge provides a naïve similarity matrix between blogs based on the tags. Such a similarity matrix is augmented by using the content of the blogs and a more robust and stable clustering approach is developed using the techniques mentioned earlier.

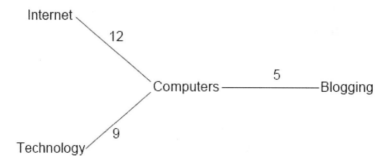

Figure 2.3: An instance of Label Relation Graph.

2.3 HYBRID APPROACH

In the previous sections, we looked at approaches that either focused on content or the graph structure to compute blog clusters and/or communities. However, clustering purely based on content misses out a whole lot of information that is encoded in the blog graph or blog network. Similarly, simply using network information for identifying blog communities misses out the valuable content information. In this section, we look at an approach that leverages both content as well as network information of the blogosphere to compute clusters and/or communities.

The essential assumption behind using content as well as network information for blog clustering or community discovery is that: a set of blogs that link more often with the blogs in this set than with those that are outside and they share similar content, reflect tighter communities or blog clusters [23]. As discussed in Section 2.1, the network information of the blogs can be represented by W. Similarly, the blogs can also be represented as a document-term matrix A as discussed in Section 2.2. A can also be the blog-tag matrix where the individual instances or the rows of the matrix A are the blogs and the features or the columns are the user-defined tags, as in [23]. A can be decomposed into reduced concept space using the SVD transformation as shown in Eq. 2.6. Then both the network and content information can be represented as:

$$W' = \begin{pmatrix} 1 & A \\ A^T & \beta W \end{pmatrix} \tag{2.7}$$

The parameter β ($0 \leq \beta \leq 1$) controls the contribution of the content matrix (A) and the link or network matrix (W) in computing the blog communities or clusters. $\beta = 1$ means equal contribution of both content as well as network information for clustering or community discovery. $\beta = 0$ means only content information is used. This yields a bipartite graph between blogs and the terms, tags, or concepts. The eigenvector decomposition of W' gives the partition of the nodes in the blog network or the communities. Such a partition minimizes the number of edges that are cut resulting in clusters that have more links within the set than outside. Moreover, such nodes share similar content. Additional constraints can be placed such as *must-link* or *cannot-link* pairs that are derived through either domain knowledge or through the data. Such constraints have been well studied in [24] under the notions of constrained spectral clustering.

CHAPTER 3

Influence and Trust

In this chapter, we discuss two significant concepts arising from network structure, position, roles of individuals and knowledge solicitations of other members, i.e., influence and trust. Influence is a characteristic of an individual that defines the capacity of exerting some effect on other individual(s). A blogger is influential if he/she has the capacity to affect the behavior of fellow bloggers. Such bloggers are also called leaders or bellwethers of the community. Trust, on the other hand, has more to do with knowledge solicitation. It is the belief in the reliability of one's actions. People believe what a trusted individual says or does. Trust acts as lubricant that improves information flow and promotes frankness and transparency, e.g., a book review by a person with real username is more trusted. Trust can also be helpful in online discussion forums where users look for informative and trustworthy answers to their questions. Giving trustworthy recommendations could also improve the customer retention policies. An influential individual is also trustworthy; however, a trustworthy individual may not be an influential individual. These concepts will be defined formally in subsequent sections.

3.1 INFLUENCE

The advent of participatory Web applications (or Web 2.0 [1]) has turned the former mass information consumers to the present information producers [2]. Examples include blogs, wikis, social annotation and tagging, media sharing, and other such services. Blogging is becoming a popular means for mass Web users to express, communicate, share, collaborate, debate, and reflect. Blogosphere provides a conducive platform to build the *virtual communities* of special interests. We studied how communities can be extracted from Blogosphere in Chapter 2. inspires viral marketing [4], provides trend analysis and sales prediction [5, 6], aids counter-terrorism efforts [7] and acts as grassroot information sources [8].

Virtual communities bear many resemblances with physical communities. Physical communities are formed by people with similar interests, where they convene and discuss issues ranging from daily matters, politics, social events to business ideas and decision. Bloggers meet in their virtual communities in Blogosphere and conduct similar activities as their counterparts in a physical community. In a physical world, according to [25], 83% of people prefer consulting family, friends or an expert over traditional advertising before trying a new restaurant, 71% of people do the same before buying a prescription drug or visiting a place, and 61% of people talk to family, friends or an expert before watching a movie. In short, before people buy or make decisions, they talk and listen to other's experience, opinions, and suggestions. The latter affect the former in their decision making, and are aptly termed as the *influentials* [25]. Influence has always been a topic of unabated interest in business and society. Influential members of a physical community have been recognized as market-movers, those who can sway opinions, affect many on a wide spectrum of decisions. For about every 10 people, one leads the other 9 [25]. With the pervasive presence and ease of use of the Web, an increasing number of people with different backgrounds flock to the Web - a virtual world to conduct many previously inconceivable activities from shopping, to making friends, and to publishing. As we draw parallels between physical and virtual communities, we are intrigued by the questions like whether there exist the influentials in a virtual community (a blog), who they are, and how to find them.

Blogs can be categorized into two major types: *individual* and *community* blogs. Individual blogs are single-authored who record their thoughts, express their opinions, and offer suggestions or ideas. Others can comment on a blog post, but cannot start a new line of blog posts. These are more like diary entries or personal experiences. Examples of individual blogs are Sifry's Alerts: David Sifry's musings (http://www.sifry.com/alerts/) (Founder & CEO, Technorati), Ratcliffe Blog – Mitch's Open Notebook (http://www.ratcliffeblog.com/), The Webquarters (http://webquarters.blogspot.com/), etc. A community blog is where each blogger can not only comment on some blog posts, but also start some topic lines. It is a place where a group or community of bloggers voluntarily get together with common interests, share their thoughts, exchange ideas and opinions, and discuss various issues related to a special interest. Examples of community blog sites are Google's Official Blog (http://googleblog.blogspot.com/), The Unofficial Apple

Weblog (http://www.tuaw.com/), Boing Boing: A Directory of Wonderful Things (http://boingboing.net/), etc. For an individual blog, the host is the only one who initiates and leads the discussions and thus is naturally the influential blogger of his/her site. Perhaps, community discovery approaches studied in Chapter 2 can be used to extract communities among a set of individual blogs and then influential blog/blogger can be identified for synthesized communities. For a community blog, many have equal opportunities to participate; it is natural to ask who are influential bloggers.

The identification of the influential bloggers can help develop innovative business opportunities, forge political agendas, discuss social and societal issues, and lead to many interesting applications. For example, the influentials are often *market-movers*. Since they can influence buying decisions of the fellow bloggers, identifying them can help companies better understand the key concerns and new trends about products interesting to them, and smartly affect them with additional information and consultation to turn them into unofficial spokesmen. As reported in [26], approximately 64% advertising companies have acknowledged this phenomenon and are starting blog advertising.

The influentials could also *sway* opinions in political campaigns, elections, and affect reactions to government policies [27]. Tapping on the influentials can help understand the changing interests, foresee potential pitfalls and likely gains, and adapt plans timely and pro-actively (not just reactively). The influentials can also help in customer support and troubleshooting since their solutions are trustworthy because of the sense of authority these influentials possess. An increasing number of companies these days host blog sites for their customers where people can discuss issues related to a product. For example, Macromedia (http://weblogs.macromedia.com/) aggregates, categorizes and searches the blog posts of 500 people who write about Macromedia's technology. Instead of going through every blog post, an excellent entry point is to start with the influentials' posts.

Some recent numbers from Technorati (http://www.technorati.com/) show a 100% increase in the size of Blogosphere every six months, ". . . , about 1.6 Million postings per day, or about 18.6 posts per second" (http://www.sifry.com/alerts/archives /000436.html). Blogosphere has grown over 60 times during the past three years. With such a phenomenal growth, novel ways have to be developed in order to keep track of the developments in the blogosphere.

The problem of ranking blog sites or bloggers differs from that of finding authoritative webpages using algorithms like PageRank [28] and HITS [29]. PageRank would assign a numerical weight for each blog post to "measure" its relative importance. The PageRank score of a blog post (p_i) is a probability $(PR(p_i))$ that represents the likelihood of a random surfer clicking on links will arrive on this blog post and is represented as:

$$PR(p_i) = \frac{1-d}{N} + d \sum_{p_j \in M(p_i)} \frac{PR(p_j)}{L(p_j)} \tag{3.1}$$

where d is the damping factor that the random surfer stops clicking at some time, $M(p_i)$ is the set of all the blog posts that link to p_i, $L(p_j)$ is the total number of outbound links on blog post p_j, and N is the total number of blog posts. The PageRank values $\mathbf{R}$ could be computed as the entries of the dominant eigenvector of the modified adjacency matrix,

$$\mathbf{R} = \begin{bmatrix} (1-d)/N \\ (1-d)/N \\ \vdots \\ (1-d)/N \end{bmatrix} + d \begin{bmatrix} l(p_1, p_1) & l(p_1, p_2) & \cdots & l(p_1, p_N) \\ l(p_2, p_1) & \ddots & & \vdots \\ \vdots & & l(p_i, p_j) & \\ l(p_N, p_1) & \cdots & & l(p_N, p_N) \end{bmatrix} \mathbf{R}$$

where the function $l(p_i, p_j)$ is 1 if blog post p_j links to blog post p_i, and 0 otherwise.

As pointed out in [30], *blog sites in the blogosphere are very sparsely linked and it is not suitable to rank blog sites using Web ranking algorithms*. The Random Surfer model of webpage ranking algorithms [28] does not work well for sparsely linked structures. The temporal aspect is most significant in blog domain. While a webpage may gain authority over time (as its adjacency matrix gets denser), the older a blog post gets the less attention it gets, and hence its influence diminishes over time. This is due to the fact that the adjacency matrix of blogs (considered as a graph) will get sparser as thousands of new sparsely-linked blog posts appear every day.

Influential Blog Sites vs. Influential Bloggers: Researchers have studied the influence in the blogosphere from the perspective of both influential blog sites as well as influential bloggers. Finding *influential blog sites* in the blogosphere is an important research problem, which studies how some blog sites influence the external world and within the blogosphere [31]. This line of research delves into identifying those blog sites that are most popular. Such blog sites could be maintained by several bloggers and little or nothing is known about the influence of an individual blogger. It is, however, different than the problem of identifying influential bloggers in a community. Regardless of a blog being influential or not, it can have its influential bloggers. Blogosphere follows a power law distribution [32] with very few influential blog sites forming the short head of the distribution and a large number of non-influential sites form the long tail where abundant new business, marketing, and development opportunities can be explored [33]. Identifying influential bloggers at a blog site is regardless of the site being influential or not.

3.1.1 GRAPH BASED APPROACH

Treating the blogosphere as a graph is a natural choice. As mentioned in Chapter 1 the blogosphere can be depicted as either a blog graph or a blog post graph. One can easily derive a blogger graph based on the blog graph. Identifying central or influential nodes in such a graph representation is a well studied problem from the social network analysis perspective. Based on the graph one can identify influential blogs, influential blog posts, or influential bloggers respectively for blog graph, blog post graph or blogger graph. We will first examine some of these measures commonly known as centrality measures before embarking on identifying central or influential nodes in the blogosphere.

Centrality measures determine the relative importance of a vertex within the graph based on its position in the graph. These measures help in studying the structural attributes of nodes in a network. The location of a node in the network determines the importance, influence or prominence of a node in the network. These measures also decide the extent to which the network revolves around a node. Four widely used centrality measures in network analysis are: degree centrality, betweenness centrality, closeness centrality, and eigenvector centrality. Next, we explain each one of them in detail.

Degree Centrality: Degree centrality refers to the number of edges incident upon a node. In the case of a directed graph, degree centrality is further defined as indegree and outdegree centrality. Higher indegree centrality refers to "popularity" of a node, whereas higher outdegree centrality refers to "gregariousness" of a node. Mathematically, degree centrality of a node u is defined as,

$$C_D(u) = \frac{d(u)}{n-1} \tag{3.2}$$

where $C_D(u)$ refers to the degree centrality of the node u, $d(u)$ refers to the degree of node u or the number of edges incident upon u or the total number of ties u has, and n denotes the total number of nodes in the graph. Note that $d(u)$ is computed using Equation 1.1. In case of a directed graph, indegree and outdegree centrality can be computed as,

$$C_{inD}(u) = \frac{d_{in}(u)}{n-1} \tag{3.3}$$

$$C_{outD}(u) = \frac{d_{out}(u)}{n-1} \tag{3.4}$$

where $C_{inD}(u)$ and $C_{outD}(u)$ denote the indegree and outdegree centrality, $d_{in}(u)$ and $d_{out}(u)$ denote the indegree and outdegree of node u and are computed using Equation 1.2 and Equation 1.3, respectively.

Example 3.1. Consider the blog graph in Figure 1.1(b). $C_{inD}(B_1) = 2/3$, $C_{inD}(B_2) = 0$, $C_{inD}(B_3) = 2/3$, and $C_{inD}(B_4) = 1/3$; $C_{outD}(B_1) = 2/3$, $C_{outD}(B_2) = 1/3$, $C_{outD}(B_3) = 0$, and $C_{outD}(B_4) = 2/3$.

Higher degree means that the node has higher probability to catch whatever is flowing through the network, hence making it more prominent.

Betweenness Centrality: Betweenness centrality of a node u refers to the ratio of the number of geodesic paths between any two nodes s and t

that pass through node u to the total number of geodesic paths that may exist between s and t. Mathematically, it is defined as,

$$C_B(u) = \sum_{s \neq u \neq t \in V} \frac{\sigma_{st}(u)}{\sigma_{st}} \qquad (3.5)$$

where $C_B(u)$ denotes the betweenness centrality of u, σ_{st} the total number of geodesic paths between s and t, $\sigma_{st}(u)$ the number of geodesic paths between s and t that pass through u, and V the set of vertices in the graph. This measure evaluates how well a node can act as a "bridge" or intermediary between different subgraphs. A node with high betweenness centrality can become a "broker" between different subgraphs.

Closeness Centrality: Closeness centrality refers to the mean geodesic distance of a node to all other nodes in the graph. The nodes that have shorter geodesic distance to other nodes in the graph have higher closeness. Mathematically, closeness centrality is defined as,

$$C_C(u) = \frac{\sum_{t \in V \setminus v} d_G(u, t)}{n - 1} \qquad (3.6)$$

where $C_C(u)$ refers to the closeness centrality of the node u, $d_G(u, t)$ denotes the geodesic path between u and t, and n is the total number of vertices in the graph with V vertices. A node with the highest closeness centrality value could be imagined as the "nearest" node to the other nodes in the network.

Eigenvector Centrality: Eigenvector centrality defines a node to be central if it is connected to those who are central. This could be gauged as the "authoritativeness" of a node. Essentially, if a blog is connected to several "popular" or well-known blogs, then automatically its prominence is increased. This is the only centrality measure that computes the prominence of a node or the influence of a node based on the influence or prominence of the nodes it is connected to. Google's PageRank algorithm is also motivated by the eigenvector centrality. For the i-th node the centrality score is the sum of the scores of the nodes it is connected to. Mathematically,

$$p_i = \frac{1}{\kappa} \sum_{j \in M(i)} p_j = \frac{1}{\kappa} \sum_{j=1}^{n} A_{ij} p_j \qquad (3.7)$$

where $M(i)$ denotes the set of nodes p_i is connected to, A the adjacency matrix, A_{ij} is 1 if i-th node is adjacent to j-th node and 0 otherwise, κ the eigenvalues, and n is the total number of nodes. In vector notation,

$$\mathbf{P} = \frac{1}{\kappa} A \mathbf{P} \qquad (3.8)$$

or as the eigenvector equation

$$A \mathbf{P} = \kappa \mathbf{P} \qquad (3.9)$$

Hence the principal eigenvector of the adjacency matrix of the network gives the eigenvector centrality scores of the nodes in the graph.

Other measures for analyzing social networks are *clustering coefficient* (the likelihood that associates of a nodes are associates among themselves, which ensures greater cliquishness), *cohesion* (extent to which the actors are connected directly to each other), *density* (proportion of ties of a node to the total number of ties this node's friends have), *radiality* (extent to which an individual's network reaches out into the network and provides novel information), and *reach* (extent to which any member of a network can reach other members of the network).

Influential nodes in the blog graph can also be identified using the information diffusion theory. Nodes that maximize the information spread [34] can be considered as the key players or influential nodes in the graph. Two fundamental models for the information diffusion process have been considered in the literature:

- Threshold Models: Each node u in the network has a threshold that determines its tolerance or susceptibility to the infection. This threshold t_u is typically drawn from a probability distribution, $t_u \in [0, 1]$. The set of neighbors of u is defined by $\Gamma(u)$. Each neighbor v of u, i.e., $v \in \Gamma(u)$, has a nonnegative edge weight, $w_{u,v}$, and $\sum_{v \in \Gamma(u)} w_{u,v} \leq 1$. $w_{u,v}$ is determined based on the strength of the tie between u and v, which could be computed in a number of ways, such as, number of interactions u and v had. The tie strength has a direct impact on the v's ability to infect u. The larger the tie strength the more chances are that v can infect u. Those neighbors of u that are infected exert infection over u and u gets infected if, $t_u \leq \sum_{infected \; v \in \Gamma(u)} w_{u,v}$.

- Cascade Models: As the name suggests cascade models simulate a cascade effect of infection. If a social contact $v \in \Gamma(u)$ gets infected, u also gets infected with probability $p_{u,v}$ which is proportional to the edge weight, $w_{u,v}$. Cascade models can be categorized into two types: independent cascade models and generalized cascade models. In an independent cascade model the influence is independent of the history of all other node activations, i.e., v gets a single chance to infect u. If v is unsuccessful in infecting u then v never attempts to infect u. A generalized cascade model [35] generalizes the independent cascade model by relaxing the independence assumption.

Gruhl et al. [36] study information diffusion of various topics in the blogosphere from individual to individual, drawing on the theory of infectious diseases. They associate 'read' probability and 'copy' probability with each edge of the blogger graph indicating the tendency to read one's blog post and copy it, respectively. They also parameterize the stickiness of a topic which is analogous to the *virulence* of a disease. An interesting problem related to viral marketing [4, 37] is how to maximize the total influence in the network (of blog sites) by selecting a fixed number of nodes in the network. A greedy approach can be adopted to select the most influential node in each iteration after removing the selected nodes. This greedy approach outperforms PageRank, HITS and ranking by number of citations, and is robust in filtering splogs (spam blogs) [38] (More on spam blogs is discussed in Chapter 4). Leskovec et al. [39] proposed a submodularity based approach to identify the most important blogs, which outperforms the greedy approach. Nakajima et al. [40] attempts to find *agitators*, who stimulate discussions; and *summarizers*, who summarize discussions, by thread detection. Watts and Dodds [41][42] studied the "influentials hypothesis" using computer simulations of interpersonal influence processes and found that large cascades of influence are driven by a critical mass of easily influenced individuals.

However, as we mentioned before, blogosphere suffers from the challenges of link sparsity due to its casual nature. Often bloggers do not cite the source they referred to write their blog post. This creates an extremely sparse structure and provides challenges in identifying influential nodes through purely network based approaches. Researchers are exploring the possibility of constructing implicit links using content similarity and temporal dimension. In the next section, we look at approaches that focus on various statistics derived from the content on the blogs to identify influential blogs/bloggers.

3.1.2 CONTENT BASED APPROACH

Blogs, as opposed to friendship networks, contain a humongous source of textual content; they present opportunities to exploit textual content in order to identify the influentials. We will discuss some techniques, which leverage the content to derive statistics that help in identifying the influentials.

Representative blog posts are the entries that represent the theme of the blog site, and can be identified by looking at the content [43]. Such entries should not only be representative in content but also diverse so that they cover most of the topics discussed on the blog site. Specifically, the problem of identifying representative blog posts from a blog B_i can be stated as, select a subset S_i of blog posts such that $|S_i| = \min(k, N_i)$ and $S_i \subseteq B_i$, where N_i is the total number of blog posts in the blog B_i, and k is a user specified parameter that sets the number of representative entries required from a blog. For each blog post $B_{ij} \in S_i$, we make sure that they are *representative* and *diverse*.

If each blog post is represented in a vector space model, then clustering is performed that groups the blog posts in clusters. Only sufficiently large clusters are retained while other clusters are discarded as noise. A centroid for each of the retained cluster is computed as,

$$c*_p = \frac{1}{N_{c_p}} \sum_{B_{ij} \in c_p} B_{ij} \qquad (3.10)$$

where $c*_p$ is the cluster centroid of cluster c_p and $1 \leq p \leq k$, N_{c_p} the size of the cluster c_p, and B_{ij} the vector space model of the blog post j of blog i. Given the cluster centroids, the **representativeness** of a blog post B_{ij} can be computed as,

$$r(B_{ij}, B_i) = sim(B_{ij}, c*_p) \times \frac{l(B_{ij})}{\max_j l(B_{ij})} \qquad (3.11)$$

where B_{ij} is a blog post that belongs to the cluster c_p, sim a similarity function that could be computed using cosine similarity between the vector space model of B_{ij} and the cluster centroid $c*_p$ of the cluster c_p, and $l(B_{ij})$ the number of distinct words in the blog post B_{ij}. The **diversity** of the selected subset of blog posts S_i is computed as,

$$d(S_i) = \sum_{B_{ij}, B_{ik} \in S_i} dist(B_{ij}, B_{ik}) \qquad (3.12)$$

where B_{ij} and B_{ik} are the selected blog posts and belong to the subset S_i, and $dist$ a distance function that computes the distance between any two blog posts.

Based on the above measures, the task of identifying representative and diverse blog posts is reduced to evaluating the following function,

$$f(S_i, B_i) = r(S_i, B_i) + \lambda d(S_i) \qquad (3.13)$$

where $r(S_i, B_i)$ denotes the representativeness of the blog posts in the subset S_i, and $d(S_i)$ the diversity of the blog posts in the subset S_i. Finding a subset S_i such that $f(S_i, B_i)$ is maximized can be written as,

$$\max_{|S_i|=k} f(S_i, B_i) \qquad (3.14)$$

where k is the top-k blog posts that the user is interested in. Maximizing this function results in a set of blog posts S_i that are most representative and diverse. The above problem is combinatorial optimization, which is often NP-hard to find the global optima. Sub-optimal solutions can be computed for such a problem using a greedy strategy as follows: start with an empty set S_i; at each iteration add a blog post B_{ij}, such that the difference $f(S_i \cup B_{ij}) - f(S_i)$ is maximized; keep on expanding the set S_i until k blog posts are included.

3.1.3 HYBRID APPROACH

We now present a system iFinder [44] that leverages both content-driven statistics and graph information to identify influential bloggers. Some of the desirable properties of an influential blog post are summarized as follows:

An Initial Set of Intuitive Properties: According to [25], one is influential if he/she is recognized by fellow citizens, can generate follow-up activities, has novel perspectives or ideas, and is often eloquent. Below, we examine how these social gestures describing the characteristic properties of the influential can be approximated by some collectable statistics.

Recognition: An influential blog post is recognized by many. This can be equated to the case that an influential post p is referenced in many other posts. The influence of those posts that refer to p can have different impact: the more influential the referring posts are, the more influential the referred post becomes. Recognition of a blog post is measured through the inlinks (ι) to the blog post. Here ι denotes the set of blogs/blog posts that link to blog post p.

Table 3.1: Social Gestures for Identifying Influential Bloggers and their Corresponding Collectable Statistics.

Social Gesture	Collectable Statistics	Notation
Recognition	Set of Inlinks	ι
Activity Generation	Number of Comments	γ
Novelty	Set of Outlinks	θ
Eloquence	Length of the Blog Post	λ

Activity Generation: A blog post's capability of generating activity can be indirectly measured by how many comments it receives, or the amount of discussion it initiates. In other words, few or no comment suggests little interest of fellow bloggers, thus non-influential. Hence, a large number of comments (γ) indicates that the post *affects* many; such that they care to write comments, and therefore, the post can be influential. There are increasing concerns over spam comments that do not add any value to the blog posts or blogger's influence. Fighting spam is a topic of Chapter 4.

Novelty: Novel ideas exert more influence as suggested in [25]. Hence, the outlinks (θ) is an indicator of a post's novelty. If a post refers to many other blog posts or articles, it indicates that it is less likely to be novel. A blog post p is less novel if it refers to more influential blog posts than if it refers to less influential blog posts. Here θ refers to the set of blogs/blog posts that blog post p refers or links to.

Eloquence: An influential person is often eloquent [25]. This property is most difficult to approximate using statistics. Given the informal nature of the blogosphere, there is no incentive for a blogger to write a lengthy piece. Hence, a long blog post often suggests some necessity of doing so. Therefore, we use the length of a post (λ) as a heuristic measure for checking if a post is influential or not. Eloquence of a blog post could be gauged using more sophisticated linguistic based measures.

The above four form an initial set of properties possessed by an influential post. We summarize these social gestures and their corresponding collectable statistics in Table 3.1. There are certainly some other potential properties. It is also evident that each of the above four may not be sufficient on its own, and they should be used jointly in identifying influential bloggers. high θ and a poor λ could identify a "hub" blog post.

Blog-post influence can be visualized in terms of an influence graph or *i-graph* in which the influence of a blog post flows among the nodes. Each node of an i-graph represents a single blog post characterized by the four properties (or parameters): ι, θ, γ and λ. i-graph is a directed graph with ι and θ representing the incoming and outgoing influence flows of a node, respectively. Hence, if I denotes the influence of a node (or blog post p), then *InfluenceFlow* through node p is given by,

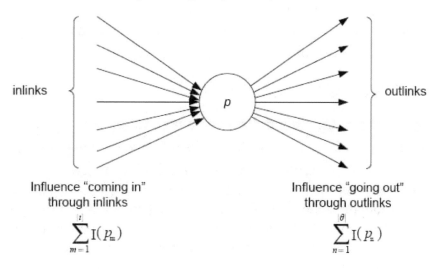

Influence "coming in"
through inlinks
$$\sum_{m=1}^{\iota} I(p_m)$$

Influence "going out"
through outlinks
$$\sum_{n=1}^{\theta} I(p_n)$$

Figure 3.1: *i-graph* showing the *Influence Flow* across blog post p.

$$Influence Flow(p) = w_{in} \sum_{m=1}^{|\iota|} I(p_m) - w_{out} \sum_{n=1}^{|\theta|} I(p_n) \qquad (3.15)$$

where w_{in} and w_{out} are the weights that can be used to adjust the contribution of incoming and outgoing influence, respectively; p_m denotes all the blog posts that link to the blog post p, where $1 \leq m \leq |\iota|$; p_n denotes all the blog posts that are referred by the blog post p, where $1 \leq n \leq |\theta|$; $|\iota|$ and $|\theta|$ are the total numbers of inlinks and outlinks of post p. *InfluenceFlow* measures the difference between the total incoming influence of all inlinks and the total outgoing influence by all outlinks of the blog post p. *InfluenceFlow* accounts for the part of influence of a blog post that depends upon inlinks and outlinks. From Eq. 3.15, it is clear that the more inlinks a blog post acquires the more recognized it is, hence the more influential it is; and an excessive number of outlinks jeopardizes the novelty of a blog post which affects its influence. We illustrate the concept of *InfluenceFlow* in the i-graph displayed in Figure 3.1. This shows an instance of the i-graph with a single blog post. Here we are measuring the *InfluenceFlow* across blog post p. Influence flows from the left (inlinks) through p to the right (outlinks).

We add up the influence "coming into" p and the influence "going out" of p and take the difference of these two quantities to get the p's influence.

As discussed earlier, the influence (I) of a blog post is also proportional to the number of comments (γ_p) posted on that blog post. We can define the influence of a blog post, p, as,

$$I(p) \propto w_{com}\gamma_p + InfluenceFlow(p) \tag{3.16}$$

where w_{com} denotes the weight that can be used to regulate the contribution of the number of comments (γ_p) toward the influence of the blog post p. We consider an additive model because an additive function is good to determine the combined value of each alternative [45]. It also supports preferential independence of all the parameters involved in the final decision. Since most decision problems like the one at hand are multi-objective, a way to evaluate trade-offs between the objectives is needed. A weighted additive function can be used for this purpose [46].

From the discussion on intuitive properties (or social gestures), we consider blog post quality as one of the parameters that may affect influence of the blog post. Although there are many measures that quantify the goodness of a blog post such as fluency, rhetoric skills, vocabulary usage, and blog content analysis, we here use the length of the blog post as a heuristic measure of the goodness of a blog post for the sake of simplicity. We define a weight function, w, which rewards or penalizes the influence score of a blog post depending on the length (λ) of the post. The weight function could be replaced with appropriate content and literary analysis tools. Combining Eq. 3.15 and Eq. 3.16, the influence of a blog post, p, can thus be defined as,

$$I(p) = w(\lambda) \times (w_{com}\gamma_p + InfluenceFlow(p)) \tag{3.17}$$

The above equation gives an influence score to each blog post. Note that the four weights (w_{in}, w_{out}, w_{com}, and w) can take more complex forms and can be tuned.

Now we consider how to use I to determine whether a blogger is influential or not. A blogger can be considered influential if he/she has at least one influential blog post. We use the blog post with maximum influence score as the representative and assign its influence score as the *blogger influence index* or *iIndex*. There could be other ways of determining the influentials. For example, if one wants to differentiate a productive influential blogger from non-prolific one, one might use another measure. For a blogger B, we can calculate the influence score for each of B's N posts and use the maximum influence score as the blogger's *iIndex*, or

$$iIndex(B) = \max(I(p_i)) \tag{3.18}$$

where $1 \leq i \leq N$. With *iIndex*, we can rank bloggers at a blog site. The top k among the total bloggers are the most influential ones. Thresholding is another way to find influential bloggers. However, determining a proper threshold is crucial to the success of such a strategy and requires more research and understanding of domain. Blog posts whose influence score is higher than the that of the top-k^{th} influential blogger could be termed as influential blog posts.

Computing Blogger Influence with Matrix Operations: We have described a hybrid model to compute the influence of a blog post using both content as well as network statistics. Here we convert the computational procedure into basic matrix operations for convenient and efficient implementation.

We define the inlinks and outlinks to the blog posts using a link adjacency matrix $\mathbf{A}$ where an entry A_{ij} is 1 if p_i links to p_j and 0 otherwise, defined as

$$A_{ij} = \begin{cases} 1 & p_i \rightarrow p_j \\ 0 & p_i \nrightarrow p_j \end{cases}$$

Matrix $\mathbf{A}$ denotes the outlinks between the blog posts. Consequently, $\mathbf{A}^T$ denotes the inlinks between the blog posts. We define the vectors for blog post length, comments, influence, and influence flow as, Now, Eq. 3.15 can be rewritten in terms of the above vectors as,

$$
\begin{aligned}
\overrightarrow{\lambda} &= (w(\lambda_{p_1}), ..., w(\lambda_{p_N}))^T, \\
\overrightarrow{\gamma} &= (\gamma_{p_1}, ..., \gamma_{p_N})^T, \\
\overrightarrow{i} &= (I(p_1), ...I(p_N))^T, \\
\overrightarrow{f} &= (f(p_1), ..., f(p_N))^T,
\end{aligned}
$$

respectively.

Now, Eq. 3.15 can be rewritten in terms of the above vectors as,

$$\vec{f} = w_{in}\mathbf{A}^T \vec{i} - w_{out}\mathbf{A}\vec{i} = (w_{in}\mathbf{A}^T - w_{out}\mathbf{A})\vec{i} \qquad (3.19)$$

and Eq. 3.17 can be rewritten as,

$$\vec{i} = \vec{\lambda}(w_c\vec{\gamma} + \vec{f}) \qquad (3.20)$$

Eq. 3.20 can be rewritten using Eq. 3.19, which can then be solved iteratively,

$$\vec{i} = \vec{\lambda}(w_c\vec{\gamma} + (w_{in}\mathbf{A}^T - w_{out}\mathbf{A})\vec{i}) \qquad (3.21)$$

The above equation requires $\mathbf{A}$ to be a stochastic matrix [47], which means all the blog posts must have at least one outlink. In other words, none of the rows in $\mathbf{A}$ has all the entries as 0. Otherwise, the influence score for such a blog post would be directly proportional to the number of comments. However, in the blogosphere, this assumption does not hold well. Blog posts are sparsely connected. This problem can be fixed by making $\mathbf{A}$ stochastic. This can be achieved by one of the following:

1. Removing those blog posts with no outlinks and the edges that point to these blog posts while computing influence scores. This does not affect the influence scores of other blog posts since the blog posts with no outlink do not contribute to the influence score of other blog posts.

2. Assigning $1/N$ in all the entries of the rows of such blog posts in $\mathbf{A}$. This implies a dummy edge with uniform probability to all the blog posts from those blog posts, which do not have a single outlink.

For a stable solution of Eq. 3.21,$\mathbf{A}$ must be aperiodic and irreducible [47]. A graph is aperiodic if all the paths leading from node i back to i have a length with highest common divisor as 1. One can only link to a blog post which has already been published and even if the blog post is modified later, the original posting date still remains the same. We use this observation to remove cycles in the blog posts by deleting those links that are part of a cycle and point to the blog posts that were posted later than the referring post. This guarantees that there would be no cycles in **A**, which makes **A** aperiodic. A graph is irreducible if there exists a path from any node to any node. Using the second strategy mentioned above by adding dummy edges to make **A** stochastic, ensures that **A** is also irreducible.

Input : Given a set of blog posts P, number of iterations *iter*, Similarity threshold τ.

Output: The influence vector, $\vec{i}$ which represents the influence scores of all the blog posts in P.

1 Compute the adjacency matrix **A**;

2 Compute vectors $\vec{\lambda}$ $\vec{\gamma}$;

3 Initialize $\vec{i} \leftarrow \vec{i_0}$;

4 **repeat**

5 $\vec{i'} = \vec{\lambda}(w_c\vec{\gamma} + (w_{in}\mathbf{A}^T - w_{out}\mathbf{A})\vec{i})$

6 *iter* $\leftarrow$ *iter* $- 1$;

7 **until** *(cosine_similarity($\vec{i}$, $\vec{i'}$ $< \tau$) V (iter ≥ 0)*;

Algorithm 2: Compute the influence scores of a set of blog posts using power iteration method.

The influence scores of blog posts can be computed by solving Eq. 3.21 using an iterative method. iFinder starts with little knowledge and at each iteration iFinder tries to improve the knowledge about the influence of the blog posts until it reaches a stable state or a fixed number of iterations specified *a priori*. The knowledge that iFinder starts with is the initialization of the vector $\vec{i}$. There are several heuristics that could be used to initialize $\vec{i}$. One way to initialize the influence score of all the blog posts is to assign each blog post uniformly a number, such as 0.5. Another way could be to use inlink and outlink counts in some linear combination as the initial values for $\vec{i}$. In iFinder, authority scores from Technorati, which are available through their API (http://technorati.com/developers/api/cosmos.html), were used. values to initialize $\vec{i}$ but since we compare our results with

PageRank algorithm we do not use it as the initial scores to maintain a fair comparison.

The computation of influence score of blog posts can be done using the well known **power iteration method** [48]. The underlying algorithm of iFinder can be described as: Given the set of blog posts P, $\{p_1, p_2, \ldots, p_N\}$, we compute the adjacency matrix $\mathbf{A}$, and vectors $\overrightarrow{\lambda}$ and $\overrightarrow{\gamma}$. The influence vector $\overrightarrow{i}$ is initialized to $\overrightarrow{i_0}$ using Technorati's authority values. Using Eq. 3.21 and $\overrightarrow{i_0}$, $\overrightarrow{i}$ is computed. At every iteration, we use the old value of $\overrightarrow{i}$ to compute the new value $\overrightarrow{i'}$. iFinder stops iterating when a stable state is reached or the user specified number of iterations is reached, whichever is earlier. The stable state is judged by the difference in $\overrightarrow{i}$ and $\overrightarrow{i'}$, measured by cosine similarity. The algorithm is presented in Algorithm 2.

3.1.4 BLOG LEADERS

Researchers have also studied various forms of blog leaders in the blogosphere. We now analyze these different types of blog leaders and compare with influential bloggers.

Based on the content type, blogs can be categorized into two classes, "Affective Blogs" and "Informative Blogs" [49]. Affective blogs are those that are more like personal accounts and diaries form of writings. Informative blogs are more technology oriented, news related, objective, and high-quality information blogs. Training a binary classifier on a hand-labeled set of blogs using Naïve Bayes, SVM, and Rocchio classifier, affective blogs can be separated from informative blogs. However, there could be influential bloggers who write affective blogs, which would be missed by such a classification.

Another type of blog leader is, who brings in new information, ideas, and opinions, then disseminate them down to the masses. This type of blog leader is known as "Opinion Leader" [50]. Their blogs are ranked using a novelty score, measured by the difference in the content of the given blog post and ones that the given blog post refers. First, the blog posts are reduced to topic space using Latent Dirichlet Allocation (LDA), and then using cosine similarity, measure between these transformed blog posts, and then using cosine similarity measure, a novelty score is computed between these transformed blog posts. Novelty score of a blog post is defined as the dissimilarity in content of the given blog post with respect to other blog posts. It is computed by averaging the cosine similarity scores of the given blog post with respect to the other blog posts. The lower the average cosine similarity score between the given blog post and the other blog posts, the higher the novelty score of the given blog post. However, opinion leaders can be different from influential bloggers. There could be a blogger who is not very novel in his/her content but attracts a lot of attention to his/her posts through comments and feedback. These bloggers will not be captured by novelty based approach. Moreover, not many blogs refer to the blogs they borrowed their content from, due to the casual nature of the blogosphere.

Many blog sites list "Active Bloggers" or top blog posts in some time frame (e.g., monthly). Those top lists are usually based on some traffic information (e.g., how many posts a blogger posted, or how many comments a blog post received) [31]. Certainly, these statistics would leave out those blog sites or bloggers who were not active. Moreover, influential bloggers are not necessarily active bloggers at a blog site.

3.2 TRUST

The past couple of years witnessed significant changes in the interactions between the individuals and groups. Individuals flock to the Internet and engage in complex social relationships, forming social networks. This has changed the paradigm of interactions and content generation. Social networking has given a humongous thrust to online communities, like Blogosphere. Trust is extremely important in social media because of its low barriers to credibility. Profiles and identities could be easily faked, and trust could be compromised, leading to severely critical physical and/or psychological losses.

Trust can be defined as the relationship of reliance between two parties or individuals. *Alice* trusts *Bob* implies *Alice*'s reliance on the actions of *Bob*, based on what they know about each other. Trust is basically prediction of an otherwise unknown decision made by a party or an individual based on the actions of another party or individual. Trust is always directional and asymmetric. *Alice* trusts *Bob* does not imply *Bob* also trusts *Alice*.

Trust can be broadly categorized into three classes [51]. When we act towards others based on the belief that their actions will suit our needs and expectations, we are indulging in anticipatory trust. The act of entrusting a valued object to a third party and expecting responsible care involves responsive trust. When we act on the belief that our trust will be reciprocated by the other person, it is a case of evocative trust. There are two major types of trust. The most commonly studied type is interpersonal trust. This involves face-to-face commitments between individuals. The second type is social trust, which involves faceless commitments towards social objects that may involve individuals in the background who are most likely unknown to us.

From a sociological perspective, trust is the measure of belief of one party in another's honesty, benevolence, and competence. Absence of any of these properties causes failure of trust. From a psychological perspective, trust can be defined as the ability of a party or an individual to influence the other. The more trusting someone is the more easily he/she can be influenced. As mentioned earlier there is a subtle difference between influence and trust. In this section we discuss what is meant by trust and related issues, e.g., how it is computed in the blogosphere, and how it propagates in the blogosphere.

3.2.1 TRUST COMPUTATION

Quantifying and computing trust in social networks is hard because concepts like trust are fuzzy, and trust is being expressed in a social way. In other words, the definitions and properties are not mathematical formulations but social ones. Due to the low barrier to publication and casual environment, there is a strict need for handling trust in social networks. However, there is limited study in computing trust in the blogosphere. Existing works like [52],[53],[54] rely on one form or the other of network centrality measures (like degree centrality, closeness centrality, betweenness centrality, eigenvector centrality) to compute nodes' trust values. Nonetheless, blog networks have very sparse trust information between different pairs of nodes because there is no explicit notion of specifying trust values to different blogs or bloggers. Although little research has been published that exploits text mining to evaluate trust in Blogosphere, authors in [55] have proposed to use sentiment analysis of the text around the links to other blogs in the network to compute trust scores.

They study the link polarity and label the sentiment as "positive", "negative", or "neutral", as illustrated in Figure 3.2. The highlighted text in the blog snippets are the links, and the underlined words/phrases denote the sentiment towards the link. Based on the words/phrases, sentiments can be identified as positive, negative, or neutral. A positive sentiment towards a link increases the trust value of the linked source, and a negative sentiment decreases the trust value of the linked source. In other words, a negative sentiment increases the distrust value of the linked source. This information mined from the blogs is coupled with Guha et al.'s [52] trust and distrust propagation approach to derive trust values between node pairs in the blog network. They further use this model to identify the distrusted nodes in the blog network and filter them out as the spam blogs. More on spam blogs will be discussed in Chapter 4.

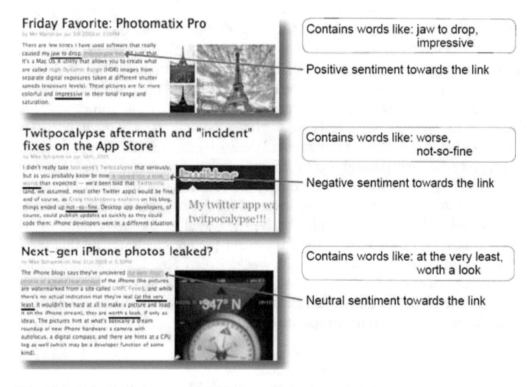

Figure 3.2: Link polarity in terms of positive, negative, and neutral sentiments.

Another approach to compute trust purely using content analysis is presented in [56]. A primary assumption of this work is that communications between individuals build trust. The more you engage your audience in interaction, the more trust you build. Various constructs are proposed to quantify the trust, including quality of effort, benevolent intent, liking, involvement, and cultural tribalism. The *quality of effort* refers to the earnest and conscientious activity intended to accomplish something. It is analyzed by the number of words used by the blogger, follow-up comments by the blogger, and number of trackbacks to the blog post. The *benevolent intent* measures the expression of good-will in the blogger's post. Harvard-Lasswell IV (H-L4) tag dictionary is used to quantify the benevolent intent from the blogger's post. The tag dictionary is a collection of words that have been pre-assigned to one or more thematic categories. One of these themes refers to the benevolent intent of the text. Depending on the number of words used by the blogger that correspond to the "benevolent" theme, we can quantify the benevolent intent from the blogger's post. *Liking* is antecedent of trust. Using the H-L4 tag dictionary, we can quantify the "liking" theme from user comments on the blogger's post. *Involvement* is a direct measure of the communications or the interactions initiated by a blogger's post. Involvement can be quantified by looking at the number of comments a blog post received, number of words in the comments, and number of unique individuals who added comments. *Cultural tribalism* directly facilitates the organizational learning, which is the process of improving actions through better knowledge and understanding. Through discussion and interaction in the blogs, the followers or readers of the blog learn and improve. It not only builds trust for the blogger but also serves as a motivation for blogging. This can be measured as the percentage of the individuals who took part in the discussion (or comments), say, this week also took part in the discussion last week. It can also be measured as the percentage of individuals who took part in the discussion this week also took part ever in the discussion before. All these constructs are quantified and a trust score is computed for the blogger.

3.2.2 TRUST PROPAGATION

Even though we assign trust scores to individual blogs or bloggers, it is still challenging to propagate these scores in the blog network. This is due to the same reason that the blog network is extremely sparse since many bloggers do not cite the source. Using trust propagation approaches for such a sparse network is extremely challenging. Note that trust is highly subjective; nevertheless, some characteristic properties are pointed:

Transitivity: Trust can propagate through different nodes following transitive property. However, the degree of trust does not remain same. It may decrease as the path length increases through which trust propagates.

Asymmetry: Trust is asymmetric, in the sense that if A trusts B then it is not necessary that B also trusts A. Some existing approaches relax this assumption and consider trust as symmetric.

Personalization: Trust is a personalized concept. Everyone has a different conception of trust with respect to some other individual. Assigning a global trust value to an individual is highly unrealistic. Trust of an individual is always evaluated with respect to some other individual.

Trust can be considered as binary-valued with 1 indicating trust and 0 indicating distrust. Trust can also be evaluated as continuous-valued. Moreover, binary-valued trust is little more complicated than meets the eye. A value of 0 could be a little vague as it could represent both no-opinion or distrust. To qualify this notion, often researchers use -1 to represent distrust and 0 as missing value or no-opinion. Researchers model the propagation of distrust the same way as the propagation of trust. Propagation of trust (T) and distrust (D) could be governed by the set of rules illustrated in Table 3.2. Here *A, B,* and *C* are different individuals and trust or distrust relationship between *A-B* and *B-C* is known. These rules help in inferring trust or distrust between *A-B*. Propagation of distrust is a little intricate. As shown in the Table 3.2, if *A* distrusts *B* and *B* distrusts *C* then *A* has reasons for either trusting *C* (enemy of enemy is a friend) or distrusting *C* (don't trust someone who is not trusted by someone you don't trust).

Table 3.2: Rules for Trust and Distrust propagation (based on [57]).

Propagation Scheme	Outcome	Comments
A $\xrightarrow{T}$ B $\xrightarrow{T}$ C	T A→C	*Transitivity*
A $\xrightarrow{T}$ B $\xrightarrow{D}$ C	D A→C	Don't trust someone who is distrusted by a person you trust.
A $\xrightarrow{D}$ B $\xrightarrow{T}$ C	D A→C	Don't trust someone who is trusted by a person whom you don't trust.
A $\xrightarrow{D}$ B $\xrightarrow{D}$ C	T (1) A→C	Enemy of your enemy is your friend.
	D (2) A→C	Don't trust someone who is not trusted by a person you don't trust.

In case the link between *A* and *C*, like *B* is missing, which can be used to infer the trust between *A-C*, a different strategy could be used. Trust only if someone is trusted by k people, i.e., if *C* is trusted by a k' number of people then *A* could trust *C*. Don't trust anyone who is distrusted by k' people, i.e. if *C* is distrusted by k' number of people then *A* could distrust *C*. Note that the thresholds k and k' could be learned from the data.

Trust is a promising area of research in social networks especially the blogosphere where most of the assumptions from friendship networks are absent.

1. Social friendship networks assume initial trust values are assigned to the nodes of the network. Unless some social networking websites allow their members to explicitly provide trust ratings for other members, it is a topic of research and exploration to compute initial trust scores for the members. Moreover, in Blogosphere it is even harder to implicitly compute initial trust scores.

2. Social friendship networks assume an explicit relationship between members of the network. However, in Blogosphere there is no concept of explicit relationship between bloggers. Many times these relationships have to be anticipated using link structure in the blogs or blogging characteristics of the bloggers such as content similarity.

3. Existing approaches for trust propagation algorithms assume an initial starting point. In Blogosphere, where both network structure and initial ratings are not explicitly defined, it is challenging to tackle the trust aspect. A potential approach could be to use influential members [44] of a blog community as the seeds for trusted nodes.

CHAPTER 4

Spam Filtering in Blogosphere

Open standards and low barriers to publication have made social media, specially Blogosphere a perfect "playground" for spammers. Spammers post non-sensical or gibberish text to many blog sites that not only degrades the quality of search results but also consumes valuable network resources. These blogs are called spam blogs or *splogs*. More formally, splogs can be defined as [58] "a blog created for any deliberate action that is meant to trigger an unjustifiably favorable relevance or importance, considering the blogs' true value". As the blogosphere continues to grow, the utility of blog search engines becomes more and more critical. They need to understand the structure of the blog sites to identify the relevant content for crawling and indexing, and more importantly, to tackle the challenging issues of splogs.

Spammers take advantage of the anonymous or guest access to post content on blog sites. Often, they create fake blogs containing either gibberish text or hijacked content from other blogs or news sources. One main objective behind this type of splogs is to host content-based advertisements, which would generate revenue if visitors accidentally click on the advertisements. Spammers also create fake blogs to host link farms, with the purpose of boosting the rank of the participating sites in the link farm. Another most popular way of spamming in blogs is through comments. Spammers use the feature of comments on the blogs as a way of publicizing their blogs or websites, which are otherwise irrelevant to the original blog. These are also known as the *spam comments*. This has made the life of spammers much easier. Instead of setting up a complex set of webpages that foster a link farm, spammers write a simple agent that visits random blogs and wikis and leave comments containing links to the spam blog. Filtering splogs and spam comments would help improve search quality; reduce wastage of storage space by discarding the splogs and spam comments in search engine index and crawlers; also, reduce the wastage of network resources.

Often blogs are setup to relay a short message or ping to the servers whenever they are updated with new content. These servers could be search engines that index these blogs for the most up to date results. Spinn3r (http://www.spinn3r.com) is one such search engine that receives these pings from blogs to update its search index. In April 2009, Spinn3r (http://spinn3r.com/spam-prevention) reported that they receive nearly 100, 000 pings per second from spammers, which forms 93% of the total pings received. This indicates the phenomenal growth of splogs. In a similar report by Akismet (http://akismet.com/stats/), the blogosphere was reported to have 10, 085, 056, 032 total spam comments till March 2009, out of which only 2, 005, 536, 845 were legitimate comments, i.e., 80.115% of all comments are spam. The phenomenal growth of spam comments is presented in the Figure 4.1(a) and (b). Figure 4.1(a) depicts the daily count of spam and legitimate comments as recorded by Akismet's (http://akismet.com/stats/) servers. Figure 4.1(b) depicts the cumulative or the total number of spam comments received till March 2009. This clearly shows an exponential growth in the number of spam comments in the blogosphere.

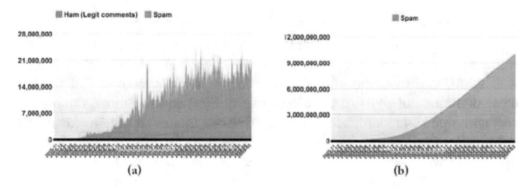

Figure 4.1: Numbers of spam comments tracked by Akismet (http://akismet.com/stats/) till March 2009 (a) daily count, and (b) total or cumulative. Copyright © Automattic, Inc. Used with permission.

Such a high volume of splogs and spam comments presents tremendous challenges in finding the legitimate blog posts and in blog search. Blog search engines are not only faced with overwhelming blog data but most of which is spam, which exacerbates the problem of efficiently finding the relevant and accurate search results. Furthermore, splogs often contain legitimate text scraped from normal blogs or news sources. Such splogs evade most of the existing spam filtering techniques that lookout for gibberish or random sequence of characters as text.

A *spam filter* is a program that keeps the legitimate content in and filters out the illegitimate or spurious content. In this case, the illegitimate content corresponds to the splogs. Spam filtering can be considered as a classic supervised learning problem with binary class labels. Splogs are treated as the positive class and non-spam blogs are the negative class. A spam filter is thus a classifier that is trained to predict the class labels for the blogs as positive (splog) or negative (non-spam blog). A correct classification could belong to either of the two categories: true positive (a splog, which is correctly classified as splog) and true negative (a non-spam blog, which is correctly classified as non-spam blog). A misclassification could also belong to either of the two categories: false positive (a non-spam blog, which is classified as splog) and false negative (a splog, which is classified as a non-spam blog). This is discussed in more detail in Chapter 5. A classification error could be disastrous. Where false positives can seriously affect the blogger's recognition and reputation, false negatives would hurt the search engine's efficiency and accuracy. So there is a pressing need to reduce and balance the classification errors.

There has been a considerable amount of work that deals with the web spam [59]. Let's look at the differences between splogs and the typical web spam, and examine why the approaches widely studied and used in these domains fall short in the blogosphere. First and foremost, the ease of use and open standards of the blogosphere makes it much easier for spammers to generate splogs and spam comments as opposed to web spam. Spammers write agents that create splogs or visit popular blogs randomly and leave spam comments. Whereas in the case of Web spam, they have to create a complex set of webpages that link to the spam webpage. Webpages do not have an option for leaving comments so they do not deal with the issues of spam comments. Second, the dynamic environment and phenomenal growth of the blogosphere defeats the spam filters that lookout for specific keywords since the splogs keep evolving and often they contain seemingly legitimate but irrelevant text. Webpages are relatively more static as compared to blogs so the content on webpages do not evolve that much. Third, due to the casual environment of the blogosphere, bloggers often use colloquial language, slangs, misspellings, and/or acronyms in their blogs. Traditional web spam filters might mistake "ham" as spam leading to a lot of false positives. These differences warrant a special treatment for splogs and spam comments that we discuss next. In this chapter, we will look at some specialized spam filtering techniques that remove splogs and spam comments using network or link and/or content.

4.1 GRAPH BASED APPROACH

Blogosphere can be represented as a network or graph of blogs as nodes and hyperlinks as vertices. We looked at an example of such a blog network in Chapter 1 in Figure 1.1(b). Given the blog network, we can study various statistics like degree distribution, clustering coefficient, and many others, as mentioned in Chapter 1. It has been shown that these statistics are considerably different for splogs and legitimate blogs [60]. These differences could be leveraged to identify splogs. For instance, the indegree for a splog does not follow densification law, i.e., over time the increase in inlinks for splogs would observe a sudden decrease [61]. Such indicators can empirically help identify splogs.

A more sophisticated measure to identify splogs leverages the correlation of increment in indegree of blogs and the blog's popularity on a search engine. The assumption here is if a blog is returned among the top search results by a search engine and if it is a legitimate blog, then it will likely attract more inlinks. If the blog was a spam and it somehow managed to get into the top results returned by the search engine through link farming or other tactics, there would be very little or no increase in the inlinks of the splog.

The above mentioned statistics could be studied at a finer granularity by considering the blog post network as mentioned in Figure 1.1(a) of Chapter 1. Each blog post is assigned a spam score based on obtained statistics. The blog is then considered as a sequence of spam scores assigned for the comprised blog posts. These spam scores are aggregated using any of the available merge functions such as: *average, max, min*, etc. To counterattack the evolving tactics used by the spammers, we can retain or place higher weights on the spam score of the more recent blog posts. The spam scores from the blog posts are combined with the above statistics for a more reliable splog filtering algorithm. However, blogs are usually sparsely linked, which presents challenges to those degree-based approaches. Some approaches attempt to densify the link structure by inferring implicit links based on the similarity in content, but more work is needed to solve various challenges considering the information quality aspects of the blogosphere due to the casual environment.

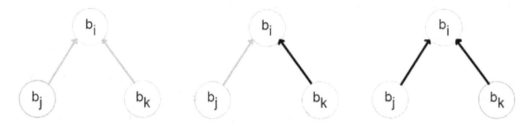

Figure 4.2: Some common network motifs of a blogger triad that accounted for spam comments.

As mentioned before, the blogosphere suffers not only from the splog issue but also from spam comments. Spammers exploit the unique feature of blogs to leave comments anonymously to gain unfair advantages. They can post links in their comments pointing to their blogs that are irrelevant to the discussion, creating link spam. Their intent is to advertise or create link farms pointing to splogs, hence promoting their blogs or webpages on search engines. Since comments usually do not have much text, content-based spam filtering approaches are likely to perform poorly in spam comments identification, resulting in both false positives and false negatives. A promising approach to identify spam comments is proposed in [62] that generates a blogger network based on the bloggers' commenting behavior. Network motifs that characterize spam commenting behaviors are then identified and used to distinguish spam blogger from a regular blogger.

A blogger network based on commenting behavior can be generated using the following procedure: consider two bloggers b_i and b_j. If b_j posts a comment on b_i's blog post that has a link pointing to b_j's blog post or blog then an edge from b_i to b_j is constructed. This network could be a weighted graph where the weight on an edge denotes the number of such comments. The study is performed on a Persian blog (http://www.persianblog.com). The blogs along with the blog posts and the comments are crawled. 700 of the comments are annotated manually using 4 human evaluators with interagreement between them being 0.96. The comments were annotated as "positive", "negative", "neutral", and "spam". Positive comments are those that encourage or support the blogger's views in the blog post. Negative comments have an opposing response to the views expressed in the blog post. Neutral comments are those that have no special sentiment towards the blog post. Spam comments are those that are irrelevant comments with an invitation to the commenter's blog/blog post or webpage. The first three classes of the comments correspond to non-spam comments.

After constructing the blogger network as described above, triads of bloggers with their commenting behavior were observed. From the observation, the most common triads that accounted for the spam comments are depicted in Figure 4.2. The dark edges denote the spam comments and lighter edges denote non-spam or regular comments. Based on this motif, a blogger b_i is highly likely (in nearly 74% cases) to be a spammer if:

1. b_i places comments on two other bloggers, b_j and b_k,

2. b_j and b_k never place comments on each other, and

3. b_j and b_k never place comments on b_i.

By studying these triads, we can determine if b_i with the relationships to b_j and b_k is a spammer or not. The computation of triads can be done very efficiently.

4.2 CONTENT BASED APPROACH

Content based splog filtering approaches treat the problem of identifying spam blogs as a binary classification task. Each blog is assigned one of the two labels: spam or ham. Based on an annotated dataset, a classifier is learned using a portion of the dataset as the training dataset. Remaining portion of the dataset is used as the test dataset to evaluate the efficiency and accuracy of the splog filtering algorithm. A blog post could be broken into various blocks: user-assigned tags, blog post content, comments, blog post title, blog post hyperlinks and anchor text. A separate classifier can be trained for each segregated block. However, this is very similar to web spam filtering approach as proposed in [63]. Due to the differences between the web spam and the blog spam mentioned earlier in this chapter, existing web spam filtering techniques do not perform well. We show how to exploit these differences and unique features of splogs to efficiently distinguish between splogs and regular blogs.

Usually splogs are machine generated pieces of text, which are either random sequence of characters that look like gibberish or blocks of legitimate text scraped from regular blogs or news sources. Content based splog filtering approaches like [58] leverage this observation to distinguish splogs from regular blogs. It has been observed that blog posts in spam blogs exhibit self-similarity over time in terms of content, links, and/or posting times. Blog posts in these splogs contain repetitive patterns in post content, affiliated links, and posting time. This peculiarity of splogs is exploited to train a splog classifier.

Self-similarity between posts i and j of a blog is defined as $S_\alpha(i, j)$, where α is one of the three attributes: content, time, and link. Each blog post i is represented as a *tf-idf* vector $\overrightarrow{h_i}$ containing the top-M terms. The content self-similarity between i and j is represented as:

$$S_c(i, j) = \frac{\sum_{k=1}^{M} \min(\overrightarrow{h_i(k)}, \overrightarrow{h_j(k)})}{\sum_{k=1}^{M} \max(\overrightarrow{h_i(k)}, \overrightarrow{h_j(k)})} \tag{4.1}$$

The link self-similarity, $S_l(i, j)$ is also computed the same way. Except constructing the *tf-idf* from terms, $S_l(i, j)$ is constructed by the domain names of the hyperlinks. Self-similarity based on time, $S_t(i, j)$ is computed by looking at the difference in the posting times as follows:

$$S_t(i, j) = \mod (|t_i - t_j|, \delta_{day}) \tag{4.2}$$

where t_i and t_j are the posting times of blog posts i and j, and δ_{day} denotes the number of time units in a day. The units are decided by the timestamp of the blog post. If the timestamp is as detailed as seconds then the time unit considered is seconds.

For a blog with N posts, a self-similarity matrix is of $N \times N$. The blog posts are arranged in increasing order of time, so a blog post i is published before $i + 1$. The self-similarity matrices thus obtained show interesting insights for splogs as follows:

Content: Splogs often have unusual topic stability while normal bloggers tend to drift over time. This is because the posts in the splogs often have repetitive blocks of text copied from other blogs or news sources.

Publication Frequencies: Normal bloggers publish blog posts regularly during few time periods, morning or night. They do not submit blog post throughout the day. However, splogs are generated by bots that are published either at same time or uniform distribution of time (throughout the day).

Link Patterns: Normal bloggers use different links as they blog on various topics. Blog posts in spam blogs often use the same or a set of links throughout.

These observations could be used to differentiate between splogs and regular blogs.

Another observation helpful to filter spam comments is the *language model disagreement*. Usually the spam comments do not link to semantically related blog posts or webpages. This divergence in the language models is exploited to effectively classify spam comments from non-spam comments [64]. Given a blog post and comments on the blog post, language models for each of them are created. The language model for each comment (Θ_1) is compared with the language model for the blog post (Θ_2) using the KL-divergence as follows:

$$KL(\Theta_1||\Theta_2) = \sum_w p(w|\Theta_1) \log \frac{p(w|\Theta_1)}{p(w|\Theta_2)} \qquad (4.3)$$

The KL-divergence gives the difference between the two distributions. The two distributions here are the language model for each comment and the language model for the blog post. If the KL-divergence value is below a threshold then the comment is considered as a regular comment or a non-spam comment, otherwise the comment is treated as a spam comment.

Blog comments can be very short, which is often the case. This could be true for some blog posts too. This results in a very sparse language model with few words. So in order to enrich the model, to achieve more accurate estimation of the language model for both the blog post and the comment, links in the blog post and the comments could be followed and the content found on these links could be added to the existing language models. These links could be followed to a certain depth which would add more content and eventually enrich the language models. Nevertheless, this also leads to the issue of topic drift and hence the language model drift. For the case of blog posts, inlinks to the blog post could also be followed to enrich the language model for the blog post.

Table 4.1: Specialized features derived from the content appearing on the blogs.

No.	Feature Description
1.	Location Entity Ratio
2.	Person Entity Ratio
3.	Organization Entity Ratio
4.	Male Pronoun Entity Ratio
5.	Female Pronoun Entity Ratio
6.	Text Compression Ratio
7.	URL Compression Ratio
8.	Anchor Compression Ratio
9.	All URLs Character Size Ratio
10.	All Anchors Character Size Ratio
11.	Hyphens Compared with Number of URLs
12.	Unique URLs by All URLs
13.	Unique Anchors by All Anchors

A significant advantage of content-based splog filtering approaches is that they do not suffer from sparse link structure of the blogosphere. This is also the reason that existing web spam filtering approaches that work on link-based approaches [59, 65] do not perform well in the blogosphere.

4.3 HYBRID APPROACH

Looking at the qualities of individual approaches: content-based and link-based, we now study a hybrid approach that combines both as proposed in [66]. First a seed set of blogs are classified as splog or regular blogs using the content-based approach, and then the link-based approach is used to expand the seed set.

The content-based approach constructs feature vector from the text in the blog using bag of words. Other options to construct the feature vector are *tf*-*idf* encoding and *n*-grams. Additional features are tokenized anchors and tokenized URLs appearing in the blog posts. Other specialized features derived from the content of blogs are presented in Table 4.1. Features 1-5 in Table 4.1 leverage the fact that splogs mention a lot of named entities such as names of places, people, organizations, etc. Features 6-8, 12 and 13 leverage the observation that splogs contain a lot of repetitive blocks of text, same URLs, and anchor text. Features 9-10 take advantage of the observation that splogs contain an unusually high number of links and anchors. Feature 11 takes advantage of the observation that splogs contain many URLs with hyphens. These features are used in conjunction with other previously mentioned features to train a support vector machine (SVM) classifier with linear kernel. Linear kernel SVM performs better than the polynomial and RBF kernel. Also, bag of words performs better as compared to *tf*-*idf* encoding and *n*-grams in terms of classification accuracy.

Once the seed set of blogs is classified, the graph-based approach is used to expand the seed set. By looking at the hyperlink graph of the blogs and assuming that "regular blogs do not link to splogs", all those blogs linked by regular blogs are classified as regular blogs. After a blog is classified as regular, it is added to the seed set and the process continues further until no more unclassified blogs are left. The hybrid approach shows how both the content-based and graph-based approaches can be tied together to achieve splog identification.

CHAPTER 5

Data Collection and Evaluation

In the previous chapters, we discussed various concepts and research opportunities in Blogosphere. This chapter specifically focuses on equally important aspects considering data collection and evaluation. Some of the concepts are difficult to evaluate using the conventional evaluation methods and require avant-garde evaluation strategies. We will also discuss some data preprocessing techniques commonly used in blog mining.

5.1 DATA COLLECTION

Data is an essential asset for performing any sort of analysis. In this section, we will discuss ways to collect data using Application Programing Interface (API) and crawlers. We will also point out some available benchmark datasets. A brief discussion will be presented regarding data preprocessing due to the noisy nature of the blogosphere.

5.1.1 API

Several social media sites like BlogCatalog, Technorati, del.icio.us, Digg, Twitter, etc. provide APIs that can be used to download part or whole of the real-world data. REST (REpresentational State Transfer) protocol is the most widely accepted protocol for these APIs, which differs from SOAP (Simple Object Access Protocol) due to the absence of session tracking as an additional messaging layer. These are often termed as RESTful APIs. Each API request is handled as a remote procedure call on the server and returns the requested portion of the data in the desired output format. It is specified as one of the arguments in the query. Most widely used formats include XML, JSON (JavaScript Object Notation), Javascript, and PHP. Usually, there are limits to the number of times one can make an API request (examples covered later in the chapter). These limits are manifested using an API key. These keys are uniquely generated for each developer. The developer is required to obtain the API key from the website and is required as an argument in the query. Under these constraints, it is rather impossible to collect all the data stored by these sites. So the best possible solution is to consider the collected data as the representative of the population and study properties of the blogosphere mentioned in previous chapters. Next, we study some specific API examples from popular social media sites and understand the returned data.

BlogCatalog (http://www.blogcatalog.com/api/) is a social community for bloggers, which is one of the largest blog directories on the internet. BlogCatalog helps in searching blogs, connecting with other bloggers, and learning and promoting your blog. Bloggers can submit their blog(s) to BlogCatalog. They can specify details such as blog URL, blog language, author name, blog country, and date when the blog was submitted to BlogCatalog. They can also specify metadata for their blog like category labels, which they think are the most appropriate for their blog, user-defined tags, and description of their blog. BlogCatalog automatically crawls the most recent 5 blog post snippets. Bloggers can provide tags for these blog post snippets. Members of BlogCatalog can rate the blogs submitted by fellow bloggers on a scale of 5 with 5 being the most popular. They can also comment on the blog/blog posts submitted by the blogger. Bloggers have the option to develop a social network with other members and can specify who are their friends. They can also specify the communities he/she is a part of and his/her favorite blogs. Blogcatalog also displays a list of 10 members who recently visited his/her blog. BlogCatalog provides a richly annotated source of blog data.

BlogCatalog provides two basic functions in its API: getinfo and bloginfo. The getinfo provides information about a BlogCatalog user. It requires two input arguments: an API key (bcwsid) and the username of the blogger whose information is to be downloaded. The returned information includes blogger's real name, blogger's details, last login attempt, name(s) of his (her) blog(s), their URL(s), their BlogCatalog URL(s), hits, views, rank, language, country, blogger's friends, and related or neighboring blogs.

Example 5.1. The sample request for getinfo looks like:

```
http://api.blogcatalog.com/getinfo?bcwsid=[apikey] &
username=johndoe
```

and the sample response looks like:

```
<?xml version="1.0" encoding="utf-8"?>
<!DOCTYPE bcapi PUBLIC "-//BlogCatalog//DTD
BCAPI 0.01//EN" "http://
 api.blogcatalog.com/dtd/bcapi-001.xml">
<bcapi version="1.0">
<result>
 <user id="56848">johndoe</user>
 <realname>John Doe</realname>
  .
  .
  .
 <weblogs>
   <weblog id="4286052">
     <name>The JohnDoe Blog</name>
     <url>http://blog.johndoe.com</url>
     <bcurl>http://www.blogcatalog.com/blogs/
     the-johndoe-blog.html
     </bcurl>
    .
    .
    .
  </weblog>
 </weblogs>
 <friends>
   <user id="56401">AngeldaVinci</user>
   <user id="13309">SiteProPlus</user>
   <user id="50995">thegoodknife</user>
    .
    .
    .
 </friends>
 <neighborhoods>
   .
   .
   .
 </neighborhoods>
</result>
</bcapi>
```

The bloginfo query provides information about a blog. It requires two input arguments: an API key (bcwsid) and the BlogCatalog url in which you are interested. This request would return the name of the blog, actual URL, BlogCatalog URL, categories that this blog is listed under, user-defined tags of the blog, hits, views, rank, language, country, number of reviews, review details (including the reviewer name, rating, review date, review text), similar or neighborhood blogs, and recent viewers. Note that actual URl is different from BlogCatalog URL. For instance, an actual URL looks like, http://blog.johndoe.com/, whereas a BlogCatalog URL looks like, www.blogcatalog.com/blogs/the-johndoe-blog.html.

Example 5.2. The sample request for bloginfo looks like:

```
http://api.blogcatalog.com/bloginfo?bcwsid=
[apikey]&url= http://
www.blogcatalog.com/blogs/the-johndoe-blog.html
```

and the sample response looks like:

```xml
<?xml version="1.0" encoding="utf-8"?>
<!DOCTYPE bcapi PUBLIC "-//BlogCatalog//DTD
BCAPI 0.01//EN" "http://
 api.blogcatalog.com/dtd/bcapi-001.xml">
<bcapi version="1.0">
<result>
 <weblog id="4286052">
   <name>;The JohnDoe Blog</name>
   <url>http://blog.johndoe.com</url>
   <user id="56848">JohnDoe</user>
   <bcurl>
       http://www.blogcatalog.com/blogs/
       the-johndoe-blog.html
   </bcurl>
   <categories>
       <category>Blog Resources</category>
       <category>Blogging</category>
   </categories>
   <tags>
       <tag>annoucements</tag>
       <tag>blogcatalog</tag>
       <tag>news</tag>
   </tags>
   <review-count>4</review-count>
   <reviews>
       <review id="78005">
           <rating>10</rating>
           <realname>Jonathan-C. Phillips</realname>
           <text-excerpt>this
           is so cool. . .</text-excerpt>
           .
           .
           .
       </review>
       .
       .
       .
   </reviews>
   <neighborhood>
       .
       .
       .
   </neighborhood>
   <recent-viewers
       .
       .
       .
   </recent-viewers
</result>
</bcapi>
```

More examples of various APIs available at different blog sites such as Technorati, Digg, del.icio.us, and Twitter are discussed in Appendix B.

5.1.2 WEB CRAWLER

Some blogs do not provide APIs to download data and most of them are not indexed by the social media sites including the ones mentioned above. In such cases, one could write crawlers, which are programs that automatically download webpages including blogs. A crawler can visit many webpages following the trail of hyperlinks, virtually moving from one webpage to another. The most important difference between conventional web crawlers and blog crawlers is that there is a continuous need for blog crawlers to download the most recent information as new content is added to the blogs. The conventional or web crawlers can be modified to make them work in an incremental fashion. First, we will briefly discuss how conventional or web crawlers work and then explain modifications to make them work incrementally.

A conventional or web crawler works in a sequential manner. It starts from a seed set of webpages (or URLs) and then uses the links within them to fetch other pages. This process continues until a stopping criterion is satisfied. Each time the crawler encounters a new URL, it adds the URL to a queue-based (FIFO) data structure known as "frontier". Each time the crawler finishes downloading the webpage, it extracts a URL from the frontier and starts crawling a new webpage. This ensures a breadth-first traversal of the web. The downloaded webpage is parsed and stored in a repository or a database and the new URLs are added to the frontier. The crawler either stops once the frontier is empty or some other criterion is satisfied. Some applications require the crawler to download webpages that belong to certain type or topic. Such crawlers are called *preferential crawlers*. These crawlers crawl webpages and add those URLs to the frontier that belong to the particular topic or type. Sometimes the frontier is implemented as a *priority queue* rather than a *first in first out* (FIFO) *queue* to implement preferential crawlers.

Blog crawling is very similar to webpage crawling except that blogs are updated very frequently. This requires the crawler to continuously crawl these blogs and look for updates. Such crawlers, also known as *incremental crawlers*, look for updated information on the blogs. They are implemented the same way as conventional crawlers but they check whether the blog post is already crawled before crawling it. The blog posts are arranged in reverse chronological sequence so once a blog post is reached, which is already crawled, there is no point in going beyond that blog post. The crawler downloads the blog, post pages, parses them, and stores them in a repository or database. These parsers could be written using regular expressions that check for a particular HTML pattern. Once an occurrence of such a pattern is found, the HTML code can be stripped, and the data element is stored in the database. Let's consider an example to illustrate this.

Example 5.3. HTML code snippet for a blog post title looks like as follows:
```
<div class="post">
<p class="filed-under">Filed under:
<a href="/category/iphone/">iPhone</a>,
<a href="/category/app-store/">App Store</a>,
<a href="/category/first-look/">First Look</a>,
<a href="/category/app-review/">App Review</a></p>
<h2 class="posttitle">
<a href="http://www.tuaw.com/2009/06/11/first-look
-get-home-for-iphone/"
rel="bookmark">First Look: Get Home for iPhone</a>
</h2>
<p class="byline">by <strong><a href="/bloggers/cory-
bohon/">Cory
Bohon</a></strong> on Jun 11th, 2009</p>
```

Now to extract the title, we can use the pattern <h2 class="posttitle"> [\\w\\W]*?</h2>. This would extract everything enclosed within the h2 tag, which also includes some HTML code. This code can be trimmed using the regular expression <[\\w\\W]*?> to obtain the title of the blog post, i.e., "First Look: Get Home for iPhone".

Another way to implement an incremental crawler is to parse the RSS (Real Simple Syndication) feeds. RSS feeds are well-formed XML files available at blogs. RSS feeds are automatically updated with the most recent entries or the blog posts published at the blog. The number of most recent blog posts that are listed in the RSS feeds vary for different blogs, but usually RSS feeds contain 15 most recent blog posts. After a blog is crawled using the conventional web crawler, one can subscribe to the RSS feeds and update the blog data as soon as new RSS feeds are available.

5.1.3 AVAILABLE DATASETS

Besides downloading datasets using APIs and crawling, there are datasets available for research and public usage. Some examples are:

- **Social Computing Data Repository** hosts data from a collection of over 20 different social media sites that have blogging capacity. The data can be accessed from http://socialcomputing.asu.edu/. Some of the prominent social media sites included in this repository are BlogCatalog, Twitter, MyBlogLog, Digg, StumbleUpon, del.icio.us, MySpace, LiveJournal, The Unofficial Apple Weblog (TUAW), Reddit, etc. The data is continuously crawled daily at pre-specified time using APIs, crawlers, and RSS feeds. At the time of writing the total data repository size was estimated to be around 15 GB. The repository contains various facets of blog data including blog site metadata like, user defined tags, predefined categories, blog site description; blog post level metadata like, user defined tags, date and time of posting; blog posts; blog post mood (which is defined as the blogger's emotions when he/she wrote the blog post); blogger name; blog post comments; and blogger social network.

- **Spinn3r**: The dataset, provided by Spinn3r.com [67], is a set of 44 million blog posts made between August 1st and October 1st, 2008. A post includes the text as syndicated, as well as metadata such as the blog's homepage, timestamps, etc. The data is formatted in XML and further arranged into tiers approximating to some degree of search engine ranking. The total size of the dataset is 142 GB uncompressed, (27 GB compressed). This dataset spans a number of big news events (the 2008 Olympics, both US presidential nominating conventions, the beginnings of the financial crisis, etc.) as well as many others you might expect to find in blog posts.

- **Nielsen Buzzmetric Dataset**: This dataset (http://www.icwsm.org/format.txt) contains roughly 14 million blog posts from 3 million blog sites collected by Nielsen Buzzmetrics in May 2006. The dataset contains about 1.7 million blog-blog links. The blog posts are written in several languages, with its majority (51%) being in English. The complete dataset is over 10 GB. The marked-up fields include: date of posting, time of posting, author name, title of the post, weblog url, permalink, tags/categories, and outlinks classified by type. However, half of the blog outlinks are missing. The data is in XML format.

- **TREC Blog Dataset**: TREC blog dataset (http://trec.nist.gov/data/blog.html) contains a crawl of 100,649 feeds during late 2007 and early 2008. The feeds were polled once a week for 11 weeks. A total number of collected feeds is: 753,681.10,615 feeds were collected daily on average. The size of the dataset is 38.6 GB uncompressed, and 8 GB compressed. It also contains reasonably sized spam.

5.1.4 DATA PREPROCESSING

Given a dataset, an essential step before any data mining task is to preprocess the data and remove the irrelevant part (e.g., noise) as much as we can. Here we will present some basic data preprocessing approaches that help improve quality.

- **Stopwords** are the commonly occurring strings that have meaning in a specific language or they can be a token that does not have linguistic meaning. For example, in the English language, words such as "a", "and", "is", "the", etc. are termed as stopwords. These words do not contribute significant information in data mining or information retrieval tasks. There is no standard list of stopwords. Often it is controlled by human input. Although some lists can be found at http://www.textfixer.com/resources/common-english-words.txt, or http://armandbrahaj.blog.al/2009/04/14/list-of-english-stop-words/. Stopword elimination from the dataset is one of the essential steps in data preprocessing and helps in drastically reducing the noise.

- **Stemming** is the process of reducing the words to their stem or root form. The process of stemming is often known as conflation. For example, a stemming algorithm would reduce words like "fisher", "fishing", "fish", "fished" to the word "fish". It is not essential for a stemming algorithm to reduce a word to its morphological root; however, it is of utter importance that all the inflected or derived words map to the same stem. Porter stemmer is one of the widely used stemming algorithms (http://tartarus.org/$\sim$martin/PorterStemmer/). Stemming helps a lot in reducing the dimensionality of text contained in blog posts.

- **Colloquial Language Usage** - Often due to the casual nature of the blogosphere, bloggers use colloquial forms of language. Apparently features that seem noisy might be informative. Services like UrbanDictionary (http://www.urbandictionary.com) can be used to unravel the informative content in the slangs, abbreviations, and/or colloquial forms of language used by the bloggers.

- **Feature Ranking** - Considering the highly textual nature of blogosphere, it is often desirable to reduce the dimensionality. Typically the number of commonly used words from English vocabulary ranges approximately 30K, but a single blog cannot use all the words. So term-vector representation of blogs could be extremely sparse and leads to the curse of dimensionality. Researchers often employ term ranking approaches like *tf-idf* [68], feature selection [69], and feature reduction approaches such as latent semantic indexing (LSI) [70], probabilistic latent semantic indexing (PLSI) [71], and latent dirichlet allocation (LDA) [72]. Such techniques help in alleviating the effect of the curse of dimensionality, enhancing generalization capability, speeding up learning process, and improving model interpretability. The feature selection approaches pick the features that are most informative using measures like information gain, minimum description length, etc. The feature reduction approaches transform the feature space from term space to concept space, hence reduced dimensionality.

In a casual environment like the blogosphere that nurtures sentiments, expressions, and emotions through writing, it is much more prevalent to observe intentionally modified spellings such as "what's uppppp?" and "this is so cooooool. .". These instances demonstrate examples of intonation[1] in written texts. These examples through misspellings clearly emphasize stress on the emotions and convey more information than the regular text. It would be undesirable to consider them as sheer misspellings and replace them with the correct spellings. To the best of our knowledge, till the writing of this book none of the text analysis approaches consider this phenomenon. But clearly it has a tremendous potential and presents a great promise for further exploration.

5.2 EVALUATION

In order to measure the difference an algorithm makes over the existing counterparts, it is necessary to systematically evaluate the merit, worthiness, and significance of the algorithm. This process is called evaluation. The algorithms or the proposed techniques are evaluated using a set of standard criteria. However, sometimes the existing criteria could not be used in evaluating algorithms or techniques in the context of the blogosphere. For instance, evaluating concepts like influence presents a big challenge due to the absence of ground truth. Evaluation models based on training and testing data fail in such situations. Performing human evaluations through surveys looks like the only solution for this challenge. However, human evaluation presents bigger challenges such as funding and recruiting unbiased yet representative users. High costs and long time are another constraints.

In this section, we will discuss some standard criteria that can be used to evaluate approaches or techniques in the blogosphere research. We will also discuss some novel and avant-garde evaluation strategies for evaluating concepts like influence, where conventional evaluation criteria fall short.

5.2.1 BLOG MODELING

A predominant way to evaluate a proposed model for blog dataset is to compare the aggregate statistics of the proposed model with the actual dataset. Some of these statistics also described in Chapter 1 include average degree, indegree and outdegree distribution, degree correlation, diameter, largest weakly connected component size, largest strongly connected component size, clustering coefficient, average path length, reciprocity, etc. The data can be divided into training and testing parts. Using the training dataset, we can learn the model parameters through maximum likelihood estimation and observe the goodness of fit of the model on the test dataset in terms of the statistics mentioned above.

Example 5.4. Using Albert-Barabasi's scale-free network model, we generated a 25,000 node network. The model started with 2 nodes and with minimum node degree 60. New nodes were added to the model until we had 25,000 nodes in the network. Now to test whether it simulates the blog network, we compare the aggregate statistics in Table 5.1. The statistics for the blog dataset were derived from the blog network crawled from Blogcatalog.com. The crawled network comprised of 23,566 nodes.

Table 5.1: Comparison of aggregate statistics for a blog network crawled from Blogcatalog.com and a scale-free network generated using Albert-Barabasi model.

Statistics	Blogcatalog	Albert-Barabasi's scale-free network
Number of Nodes	23,566	25,000
Number of Node-Node Links	1,165,622	2,967,728
Scaling Exponent (degree distribution)	1.169	1.691
Average Degree of Nodes	98	118
Diameter of the Network	5	4
Clustering Coefficient	0.51	0.291
Average Shortest Path Length	2.379	2.454

5.2.2 BLOG CLUSTERING AND COMMUNITY DISCOVERY

Since clustering and community discovery are quite close to each other in terms of the homophily assumption, their evaluation is also done in a similar way. The underlying assumption for both clustering and community discovery is that similar or like-minded individuals interact more frequently as opposed to others. This phenomenon is used to evaluate the performance of clustering algorithms or community extraction algorithms. Essentially, we look at within cluster and between cluster distance. An algorithm that gives the smallest within-cluster distance and largest between-cluster distance is considered to be the best. A small within-cluster distance ensures cohesive clusters and a large between-cluster distance ensures that dissimilar clusters are well separated. For multiple clusters, we compute the average within-cluster distance (W_d) and average between-cluster distance (B_d). The ratio $\sigma = W_d/B_d$ is computed for different algorithms and the algorithm whose ratio is minimum is chosen as the best algorithm. σ can also be computed for community discovery algorithms, and this measure can also be used to evaluate the community discovery algorithms. σ is defined as:

$$\sigma = \frac{\frac{1}{k}\sum_{c_i}\left(\frac{2}{\|c_i\|\times(\|c_i\|-1)}\sum_{v_m\in c_i, v_n\in c_i} d(v_m, v_n)\right)}{\frac{2}{k(k-1)}\sum_{c_i,c_j,i<j}\left(\frac{1}{\|c_i\|\times\|c_j\|}\sum_{v_m\in c_i}\sum_{v_n\in c_j} d(v_m, v_n)\right)} \qquad (5.1)$$

$$s.t. \ 2 \leq k \leq \|D\|$$

In the above formula, c_i, c_j represent two different clusters i and j; v_m, v_n are two different vectors representing two different bloggers; k, the number of clusters, varies from 2 to the total number of blogs, i.e., $\|D\|$; and $d(v_m, v_n)$ gives the distance between two bloggers m and n, represented by the corresponding vectors v_m and v_n. Distance between two vectors can be computed as the inverse of the cosine similarity between the two vectors.

An important point is the between-cluster distance calculation. There are several ways of distance computation:

- Cluster Mean/Median Distance: Distance between different clusters could be computed using the cluster centroids or the cluster median. For clusters, c_i and c_j, their cluster mean or median could be computed as $\overline{c_j}$ and $\overline{c_i}$. The distance between these cluster means or medians is considered as the distance between two clusters. The distance between cluster medians is resistant to noise or cluster outliers whereas the distance between cluster means or centroids is not. However, this approach to compute distance requires computing the mean or median of the clusters.

- Single Linkage: Distance between two clusters, c_i and c_j is considered as the shortest pairwise distance between the cluster members of c_i and c_j. Single linkage is highly sensitive to noise or outliers. This scheme suffers from the *chaining phenomenon*, which means that clusters may be considered close to each other due to single elements being close to each other, even though many of the elements in each cluster may be very distant to each other.

- Complete Linkage: Distance between two clusters, c_i and c_j is considered as the largest pairwise distance between the cluster members of c_i and c_j. Complete linkage is highly sensitive to noise or outliers. Two clusters may be considered distant from each other due to single elements being far from each other, even though many of the elements in each cluster may be very close to each other

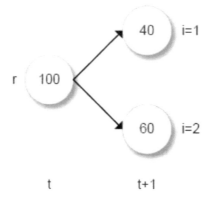

Figure 5.1: An example of cluster evolution. A cluster r consisting of 100 members at time t splits into 2 clusters, $i = 1$ and $i = 2$, consisting of 40 and 60 members, respectively, at time $t + 1$.

- Average Linkage: Distance between two clusters c_i and c_j is considered as the average pairwise distance between the cluster members of c_i and c_j. Average linkage is resistant to noise or cluster outliers.

Some clustering algorithms study the evolution in the clusters in the blogosphere. As mentioned in Chapter 2, the clusters are dynamic, which means that bloggers' interests drift over time that leads to their change in the membership of the clusters. To measure this drift of the bloggers' interests, authors in [74] propose a measure called *blogger entropy*. It measures the dispersion of the bloggers in one cluster throughout the clusters of the next time window. For a fixed value of k, i.e., the number of clusters, if many bloggers in a cluster at time t still remain in the same cluster in time $t + 1$, then the blogger entropy approaches 0. Conversely, if the bloggers in a cluster at time t are spread equally among many clusters at time $t + 1$, blogger entropy approaches 1. Blogger entropy U_r for a cluster r can be computed as:

$$U_r = -\frac{1}{\log q} \sum_{i=1}^{q} \frac{n_r^i}{n_r} \log \frac{n_r^i}{n_r} \tag{5.2}$$

where n_r is the number of bloggers that are in the dataset at time $t + 1$ and were members of the cluster r at time t, n_r^i is the number of bloggers from cluster r at time t contained in cluster i at time $t + 1$, and q is the number of clusters at time $t + 1$ among which the bloggers of cluster r at time t are distributed.

Example 5.5. To illustrate the above measure for evolutionary clustering, we study an example of a cluster r with 100 members at time t splitting in two clusters $i = 1$ and $i = 2$ of 40 and 60 members, respectively, at time $t + 1$. For the sake of simplicity, we assume that no member leaves and no new member joins. This is illustrated in Figure 5.1. Here $n_r = 100$, $n_r^1 = 40$, and $n_r^2 = 60$. So,

$$
\begin{aligned}
U_r &= -\frac{1}{\log_2 2}\left(\sum_{i=1}^{2} \frac{n_r^i}{n_r} \log_2 \frac{n_r^i}{n_r}\right) \\
&= -\frac{1}{\log_2 2}\left(\frac{n_r^1}{n_r} \log_2 \frac{n_r^1}{n_r} + \frac{n_r^2}{n_r} \log_2 \frac{n_r^2}{n_r}\right) \\
&= -\frac{1}{\log_2 2}\left(\frac{40}{100} \log_2 \frac{40}{100} + \frac{60}{100} \log_2 \frac{60}{100}\right) \\
&= 0.9709
\end{aligned}
$$

A high value of U_r (close to 1) indicates a big change in the cluster from time t to $t + 1$.

5.2.3 INFLUENCE AND TRUST

As we mentioned earlier in Chapter 3, both influence and trust are subjective concepts, which presents unique challenges in evaluation due to the absence of ground truth. In these cases, researchers often resort to a time consuming choice of human evaluation (e.g., focus groups). It is not only time consuming to acquire unbiased human evaluations but also usually expensive. An interesting alternative evaluation strategy is presented in [44]. It proposes to leverage the power of social media in evaluation. Specifically, Digg was used to evaluate how influential blog posts are.

Every story that appears on Digg is voted by the community. These votes are denoted by "diggs". The more diggs a story gets, the more popular it is. However, all posts/stories are digged or only those that leave impressions receive diggs. Hence, diggs of a post can be approximated as an influence measure. The diggs assigned to these stories are purely community contributed. These scores can be used to rank these stories based on their influence on the members of the community. This can also be treated as a survey where the human subjects voluntarily rank the stories. Using the scores on Digg, we can generate the ground truth or the gold standard against which various models can be evaluated.

Once a ground truth is obtained by either using human evaluation or social media, the evaluation of two models reduces to computing rank correlations between the two ordered lists, the one obtained by the ground truth and the other generated by the model. Note that models that compute trust scores for blog posts or blogs can also be evaluated using the measures mentioned below. Various measures like Normalized Discounted Cumulative Gain (NDCG), correlation measures such as Spearman's rank correlation and Kendall tau distance can be used. Next, we discuss these in detail.

NDCG measures the relevance of the rankings of entities such as blog posts or blogs. It allows multiple levels of relevance (r) for the entities such as irrelevant (0), borderline (1), and relevant (2). These relevance scores are defined *a priori*. We assume three levels of relevance for the sake of simplicity. The levels of relevance depend on how precisely one wants to compare the results of two algorithms.

Table 5.2: Two ranking algorithms RF_1 and RF_2 evaluated against the ground truth GT using NDCG measure as presented in Example 5.6.

i	Ground Truth		Ranking Function$_1$		Ranking Function$_2$	
	Document Order	r_i	Document Order	r_i	Document Order	r_i
1	d4	2	d3	2	d3	2
2	d3	2	d4	2	d2	1
3	d2	1	d2	1	d4	2
4	d1	0	d1	0	d1	0
	NDCG$_{GT}$=1.00		NDCG$_{RF1}$=1.00		NDCG$_{RF2}$=0.9203	

Each relevant blogs or blog post contributes some gain that is cumulated, also known as Cumulative Gain (*CG*) and is computed as:

$$CG(d_1, ..., d_n) = \sum_{i=1}^{n} r_i \tag{5.3}$$

Here $d_1, \ldots, d_n$ denote n blogs or blog posts and r_i denotes the relevance of the bog post or blog ranked at i-th position. Gains from low ranked documents are discounted, also known as Discounted Cumulative Gain (*DCG*). This ensures more relevant blogs or blog posts are ranked higher by the ranking algorithm.

$$DCG(d_1, ..., d_n) = r_1 + \sum_{i=2}^{n} \frac{r_i}{\log_2 i} \tag{5.4}$$

The final score is normalized by the maximum score possible (*MaxDCG*), finally known as Normalized Discounted Cumulative Gain (*NDCG*). Note that R denotes the ranking of the blog posts or blogs obtained through the ground truth.

$$MaxDCG = R_1 + \sum_{i=2}^{n} \frac{R_i}{\log_2 i} \tag{5.5}$$

$$NDCG(d_1, ..., d_n) = DCG(d_1, ..., d_n)/MaxDCG \tag{5.6}$$

Example 5.6. Consider the example in Table 5.2; there are 4 documents or blog posts d_1, d_2, d_3, and d_4. The ideal ranking is shown in the first column "Ground Truth". Two other ranking algorithms are denoted by RF_1 and RF_2 in the second and third columns, respectively.

$$DCG_{GT} = 2 + \{\frac{2}{\log_2 2} + \frac{1}{\log_2 3} + \frac{0}{\log_2 4}\} = 4.6309$$

$$DCG_{RF_1} = 2 + \{\frac{2}{\log_2 2} + \frac{1}{\log_2 3} + \frac{0}{\log_2 4}\} = 4.6309$$

$$DCG_{RF_2} = 2 + \{\frac{1}{\log_2 2} + \frac{2}{\log_2 3} + \frac{0}{\log_2 4}\} = 4.2619$$

$$Max DCG = DCG_{RT} = 4.6309$$

Hence, $\text{NDCG}_{RF1} = 1.00$ and $\text{NDCG}_{RF2} = 0.9203$. This exemplifies that NDCG decreases if a more relevant document is pushed down in the rank. NDCG can capture the change in ranking order.

Correlation Measures in statistics describe the relationship between two different rankings on the same set of items. A rank correlation coefficient measures the correspondence between two rankings. We discuss two most commonly used measures: Spearman's correlation coefficient (ρ) and Kendall tau distance (τ).

Spearman's correlation coefficient computes the similarity between the rankings of the corresponding items of a set. It is computed as:

$$\rho = 1 - \frac{6 \sum_i \delta_i^2}{n(n^2 - 1)} \tag{5.7}$$

where δ_i denotes the difference between the ranks of the corresponding items of the two ordered lists, and n represents the total number of items in the ordered lists. Note that Spearman's correlation coefficient compares two ranked lists of the same size with the same items. ρ lies in the interval $[-1, 1]$. $\rho = 1$ implies a perfect agreement while $\rho = -1$ implies a perfect disagreement. We illustrate the computation of Spearman's correlation coefficient with an example below.

Table 5.3: Conversion of documents to their rankings based on ground truth *GT* and ranking algorithms RF_1 and RF_2.

	d_4	d_3	d_2	d_1
GT	1	2	3	4
RF_1	2	1	3	4
RF_2	3	1	2	4
$\delta_i(RF_1)$	1	-1	0	0
$\delta_i(RF_2)$	2	-1	-1	0
$\delta_i(RF_1)^2$	1	1	0	0
$\delta_i(RF_2)^2$	4	1	1	0

Example 5.7. Continuing with the same example of the four documents d_1, d_2, d_3, and d_4 in Table 5.2. We will compute the Spearman's correlation coefficient for ranking algorithms RF_1 and RF_2 with respect to the ground truth *GT*. First we convert the document list to their rankings. The converted rankings are displayed in the Table 5.3. It shows the converted rankings for the ground truth (*GT*) and ranking algorithms RF_1 and RF_2. It also shows the difference ($\delta_i(RF_1)$ and $\delta_i(RF_2)$) in the ranking of corresponding items for RF_1 and RF_2 with respect to *GT*, respectively. Hence,

Table 5.4: Computing the discordant pairs between *GT* and RF_1; and *GT* and RF_2 for Example 5.8.

Pairs of Documents	Difference in Ranks			Discordant Pairs	
	GT	RF_1	RF_2	GT vs. RF_1	GT vs. RF_2
$d_4 - d_3$	-1	1	2	Yes	Yes
$d_4 - d_2$	-2	-1	1	No	Yes
$d_4 - d_1$	-3	-2	-1	No	No
$d_3 - d_2$	-1	-2	-1	No	No
$d_3 - d_1$	-2	-3	-3	No	No
$d_2 - d_1$	-1	-1	-2	No	No

$$\rho_{RF_1} = 1 - \frac{6 \sum_i \delta_i (RF_1)^2}{n(n^2 - 1)} = 0.8$$

$$\rho_{RF_2} = 1 - \frac{6 \sum_i \delta_i (RF_2)^2}{n(n^2 - 1)} = 0.6$$

This shows that RF_1 is a better ranking algorithm than RF_2.

Kendall tau measures the dissimilarity between two ranked lists in terms of the ranks of their corresponding items. It is computed as

$$\tau = \frac{P}{n(n-1)/2} \qquad (5.8)$$

where n is the total number of items in the ranked list, and P is the total number of discordant pairs in the two ranked lists. In order to calculate the number of discordant pairs, we pair each item with every other item and count the number of times the values in the first list are in the opposite order of the values in the second list. This is exemplified in Table 5.4. For GT and RF_1 document pair d_4 and d_3 is the only discordant pair as evident from the disagreement between the sign of difference in ranks (-1 and 1). Kendall tau distance is normalized by the total number of possible pairs $n(n - 1)/2$. The value of τ lies in the interval [0, 1]. $\tau = 0$ implies perfect agreement while $\tau = 1$ implies perfect disagreement. Let's illustrate the computation of Kendall tau distance with the help of an example.

Example 5.8. Continuing with the example of the four documents d_1, d_2, d_3 and d_4. We look at the first three rows of the Table 5.3. Table 5.4 illustrates the computation of discordant pairs between GT and RF_1 and GT and RF_2. The number of discordant pairs between ranked list generated by GT and RF_1 is 1 (counted by the number of sign disagreements between the difference in rank of the document pairs in Table 5.4). Hence $P_{RF1} = 1$. Similarly, the number of discordant pairs between ranked list generated by GT and RF_2 are 2. Hence $P_{RF2} = 2$. So,

$$\tau_{RF_1} = \frac{1}{4(4-1)/2} = 1/6 = 0.1666$$

$$\tau_{RF_2} = \frac{2}{4(4-1)/2} = 2/6 = 0.3333$$

This also shows that RF_1 is a better ranking algorithm than RF_2 since the shorter the distance the better.

Examples 5.6, 5.7, and 5.8 demonstrate NDCG's emphasis on the position of relevant documents as opposed to merely looking at their position in the case of rank correlation measures. Both Spearman's correlation coefficient and Kendall tau distance showed that RF_1 is not as good as the ground truth GT although document d_4 and d_3 have the same relevance (2). Example 5.6 shows that interchanging the position of d_4 and d_3 in the ranking would not make any difference.

5.2.4 SPAM

As we mentioned earlier in Chapter 4, spam (splog) filtering is a typical classification task, and we evaluate the various approaches based on the metrics used to evaluate classification algorithms, such as precision, recall, and f-measure. The evaluation of classification algorithms often involves a training and test model. A dataset is split into training and testing portions. An algorithm learns on the training portion of the dataset and is evaluated on the testing portion.

For the task of filtering spam, the instance of a dataset is pre-annotated as "spam" or "nonspam", "positive" or "negative", respectively. The classification algorithm treats the "spam" instances as the positive class and "non-spam" instances as the negative class. A classifier is learned using the training data and predicts the class labels of the test dataset. We can compute statistics such as *true positives, true negatives, false positives*, and *false negatives*.

True Positives (TP): Those instances that are classified as positive and actually belong to the positive class are called true positives. In terms of spam filtering, true positives are the splogs that are correctly classified as splogs by the classifier. These come under the correct classifications.

True Negatives (TN): Those instances that are classified as negative and actually belong to the negative class are called true negatives. In terms of spam filtering, true negatives are the nonspam blogs that are correctly classified as non-spam blogs by the classifier. These come under the correct classifications.

False Positives (FP): Those instances that are classified as positive but actually belong to the negative class are called false positives. In terms of spam filtering, false positives are the non-spam blogs that are misclassified as splogs by the classifier. They are a part of the misclassifications.

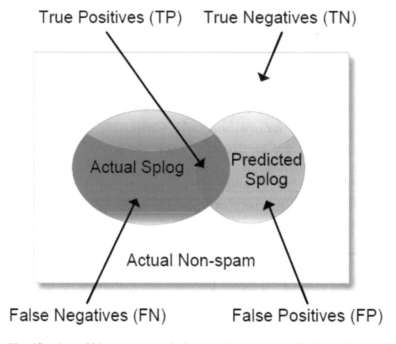

Figure 5.2: Classification of blogs as spam (splogs) and non-spam. The blue ellipse represents the set of actual splogs and the green sphere represents the predicted splogs. Everything except the blue represents the actual non-spam blogs.

False Negatives (FN): Those instances that are classified as negative but actually belong to the positive class are called false negatives. In terms of spam filtering, false negatives are the splogs that are misclassified as non-spam blogs by the classifier. They are a part of the misclassifications.

The above statistics are illustrated in a venn diagram shown in Figure 5.2. Based on these four statistics, every classifier can be evaluated using *precision* and *recall*. Precision can be defined as:

$$Precision = \frac{TP}{TP + FP} \qquad (5.9)$$

and recall is defined as:

$$Recall = \frac{TP}{TP + FN} \qquad (5.10)$$

If a classification algorithm makes a single prediction which is correct, then the precision is 100% but recall could be low. On the contrary, if the classification algorithm predicts all the instances as splogs, then the recall is 100% but precision could be low. So it is required to compare classification algorithms in terms of both precision and recall. One way is to compute the harmonic mean of precision and recall, which is commonly known as *f-measure*.

$$f\text{-}measure = \frac{2 \times Precision \times Recall}{Precision + Recall} \qquad (5.11)$$

Let's look at an example to compute the above mentioned statistics and precision, recall and f-measure for a classification algorithm.

Table 5.5: Confusion matrix for the Example 5.9.

		Actual	
		Splog	Non-spam
Predicted	Splog	7	4
	Non-spam	3	6

Example 5.9. Refer to the confusion matrix illustrated in Table 5.5. There are actually 10 splog instances and 10 non-spam blog instances, but the classification algorithm predicts 11 splog instances and 9 non-spam blog instances. Let's compute the precision, recall, and f-measure of this classifier. Here TP=7; TN=6; FP=4; FN=3. Hence,

$$Precision = \frac{TP}{TP+FP} = \frac{7}{7+4} = \frac{7}{11}$$

$$Recall = \frac{TP}{TP+FN} = \frac{7}{7+3} = \frac{7}{10}$$

$$fmeasure = \frac{2 \times Precision \times Recall}{Precision + Recall} = \frac{2 \times 7/11 \times 7/10}{7/11 + 7/10} = 0.663$$

[1]Intonation is a linguistic concept that refers to the different meanings conveyed by the different ways of pronunciation of a word [73]. The listener could interpret different meanings based on the prosodic utterance. Intonation is as common in written texts as it is in spoken language.

Tools in Blogosphere

Several modeling tools are available to simulate the social network embedded in the blogosphere that help in better understanding of various characteristics of these networks and conduct experiments. We will describe one such tool, BlogTrackers [75], in detail and briefly mention the others.

Sociologists are interested in studying the blogosphere for tracking socio-behavioral patterns, identifying the influential people in the region of interest and tracking interesting activities. They often have to eyeball the sites for useful information. Given a gamut of interests in the blogosphere, this can be a tedious and time consuming task. BlogTrackers is a user-oriented application that alleviates this problem by assisting them in effectively tracking and analyzing blogosphere. BlogTrackers grants sociologists the freedom to choose the blog sites they wish to analyze, observe interesting events and patterns with the flexibility of drilling-in. The tool consists of a number of analyzing and crawling modules and is a convenient alternative to eyeballing the blog sites and concentrate efforts on further analysis.

Most tools are generic in nature and cannot be directly used by sociologists and others with specific needs. BlogTrackers is particularly designed for their needs that can perform both data collection and provide convenient visualizing tools to analyze the data. Table A.1 presents a comparison of BlogTrackers with some of the existing tools like Technorati (http://technorati.com/), BlogPulse (http://www.blogpulse.com/), BlogScope [76] (http://www.blogscope.net/), GTD Explorer [77] (http://www.cs.umd.edu/hcil/gtd/gtd/intro.html), IceRocket (http://www.icerocket.com/), Google Blog Search (http://blogsearch.google.com/) that are specifically tailored for the blogosphere. Although, sites like Technorati and BlogPulse provide features similar to BlogTrackers, they cannot be directly used. BlogTrackers combines them in a unique manner to maximize the analytical capability of the individual techniques.

BlogTrackers is a Java based desktop application that provides a unified platform for the user to crawl and analyze blog data. It grants the user, the freedom to choose the data of interest and helps in effectively analyzing it. The data is stored in a relational database. Currently, it is tracking 21 different data sources like Twitter, Engadget, The Unofficial Apple Weblog (TUAW), LiveJournal, Flickr, Blogcatalog etc. The framework consists of two main components: Crawler and Tracker.

Crawler: BlogTrackers offers two types of crawlers to the user. The *batch crawler* crawls the websites from scratch and stores it in a database through HTML scraping using regular expressions to parse data from the HTML files. The *RSS* (Really Simple Syndicate) crawler on the other hand incrementally crawls the websites by retrieving information from their feeds. RSS crawler can be scheduled to run automatically and update the database.

Table A.1: Comparison of various analysis and visualization tools for Blogosphere (based on [75]).

	Source Selection	Data crawler	Influential Bloggers	Watch Lists/Alerts	Blog Browser	Traffic Trends	Search	Conversation tracker	Keyword Trends	Tag Clouds
BlogTrackers	✓	✓	✓	✓	✓	✓	✓	✗	✗	✓
Technorati	✓	✓	✗	✓	✓	✗	✓	✗	✓	✓
BlogPulse	✓	✓	✗	✗	✗	✗	✓	✓	✓	✓
BlogScope	✓	✓	✗	✗	✓	✓	✓	✗	✓	✓
GTD Explorer	✓	--	✗	✗	✗	✓	✓	✗	✓	✗
IceRocket	✓	✓	✗	✗	✗	✗	✓	✗	✓	✓
Google Blog Search	✓	✓	✗	✓	✓	✗	✓	✗	✗	✗

Tracker: The tracker component provides the user with a set of tools to analyze the data. The blog site to be used for analysis can be chosen by the user. The following are the 4 major tools in BlogTrackers:

1. **Blog Analysis:** BlogTrackers contains a blog browser that can be used to individually analyze the blog posts within a time period, as shown in Figure A.1. The time window can be adjusted as required. Entire blog posts are indexed for better viewing experience. Another tool for blog analysis is Term Frequency Analyzer, which shows the tag cloud (a visualization which highlights the terms by their frequency by varying the font) for all the blog posts in a given time period. This tool can be used to identify key terms associated with the blog posts during a particular time period. A traffic pattern graph can also be generated for a particular time period as shown in Figure A.2. The bar graph shows the traffic bursts depending on the granularity chosen (daily, weekly, monthly, or yearly). The bursts can be individually analyzed to observe the blog posts and the tag cloud for that period.

2. **Blogger Analysis:** BlogTrackers can be used to search for influential bloggers at a blog site. The influential bloggers are generated as described in [44]. It is also possible for a user to drill-in and look at the tags and the blog posts of the influential bloggers, as shown in Figure A.3. Bloggers can be classified based on their activity and influence into different categories like Active-Influential, Inactive-Influential, Active-Non Influential, and Inactive-Non Influential. These categories can be visualized as a confusion matrix as shown in Figure A.4.

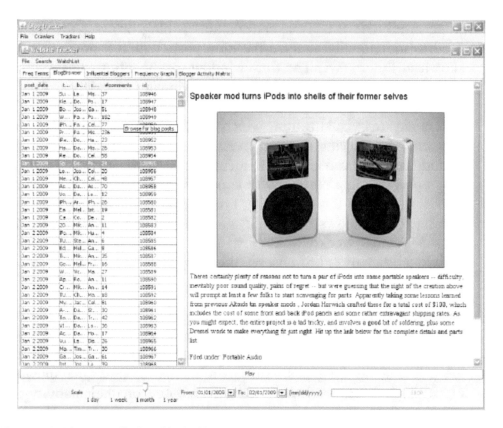

Figure A.1: Blog posts displayed in the blog browser of BlogTrackers.

3. **Search:** The blog sites are crawled on a daily basis and the posts are stored and indexed using Lucene. The index is automatically updated and can be used to search the blog posts for specific queries. The search interface of the BlogTrackers is shown in Figure A.5.

4. **Watchlists/Alerts:** BlogTrackers offers a convenient notification system to the users. A user can specify terms in the watchlist. The user is then notified by e-mail if any new post contains that word. The user can also choose the terms from a system-generated list based on popularity as shown in Figure A.6.

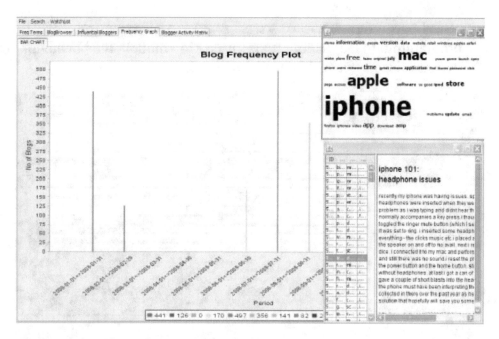

Figure A.2: Traffic pattern graph also displaying the tag cloud of the high activity period from the BlogTrackers.

Apart from the tools that are specifically tailored for Blogosphere, there are also some generic visualization software tools that do not target blogs, per se, but can be used to do some analysis on the blog data. Pajek (http://vlado.fmf.uni-lj.si/pub/networks/pajek/) is a visualization tool that can be used to visualize the network data in various ways. IBMÕs ManyEyes (http://manyeyes.alphaworks.ibm.com/manyeyes/) is another interesting project on generic visualizations but suffers from scalability issues. The Prefuse visualization toolkit (http://prefuse.org/) contains a set of unique visualizations for the data. These and many other are briefly summarized below:

- NetLogo: (http://ccl.northwestern.edu/netlogo/) A multi-agent programming language and modeling environment designed in Logo programming language. Modelers can give instructions to hundreds or thousands of concurrently operating autonomous "agents". This helps in exploring the connection between the individuals (micro-level) and the patterns that emerge from the interaction of many individuals (macro-level).

- **StarLogo:** (http://education.mit.edu/starlogo/) An extension of Logo programming language. It is used to model the behavior of decentralized systems like social networks.

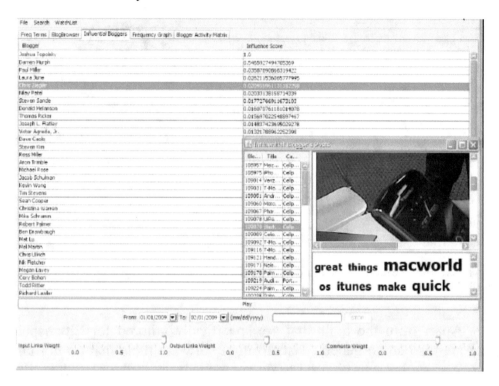

Figure A.3: Displaying the influential bloggers during a specified time period from the BlogTrackers.

- **Repast:** (http://repast.sourceforge.net/) Recursive Porous Agent Simulation Toolkit is an agent-based social network modeling toolkit. It has libraries for genetic algorithms, neural networks, etc. and allows users to dynamically access and modify agents at run time.

- **Swarm:** (http://www.swarm.org/wiki/Main_Page) A multi-agent simulation package to simulate the social or biological interaction of agents and their emergent collective behavior.

- **UCINet:** (http://www.analytictech.com/) A comprehensive package for the analysis of social network data including centrality measures, subgroup identification, role analysis, elementary graph theory, and permutation-based statistical analysis. In addition, the package has strong matrix analysis routines, such as matrix algebra and multivariate statistics.

- Pajek: (http://vlado.fmf.uni-lj.si/pub/networks/pajek/) (Slovenian: spider) A software for analyzing and visualizing large networks like social networks.

- Network package in "R": (http://cran.r-project.org/src/contrib/Descriptions/network.html) The network class can represent a range of relational data types, and support arbitrary vertex/edge/graph attributes. This is used to create and/or modify the network objects and is used for social network analysis (SNA).

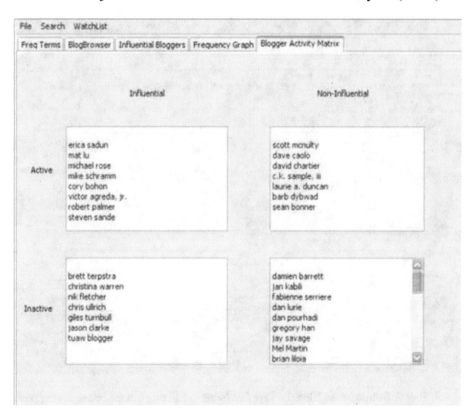

Figure A.4: Analyzing the blogger categories during a specified time period from the BlogTrackers.

- **InFlow:** (http://www.orgnet.com/inflow3.html) Another integrated product for network analysis and visualization. It has been used in the SNA domain.

- **NetMiner:** (http://www.netminer.com/) A tool for exploratory network data analysis and visualization. NetMiner allows to explore network data visually and interactively, and helps in detecting underlying patterns and structures of the network.

- **SocNetV:** (http://socnetv.sourceforge.net/) A Linux based SNA and visualizing utility. SocNetV can compute network and actor properties, such as distances, centralities, diameter, etc. Furthermore, it can create simple random networks (lattice, same degree, etc.).

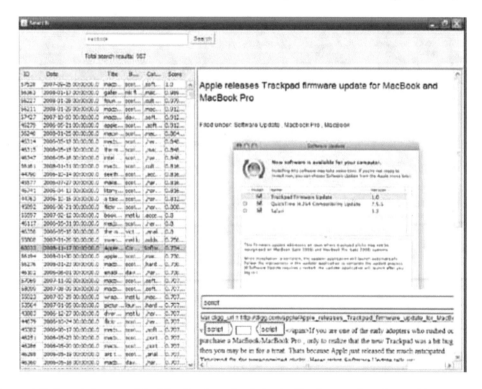

Figure A.5: Search feature of the BlogTrackers that helps in filtering out the relevant blog posts.

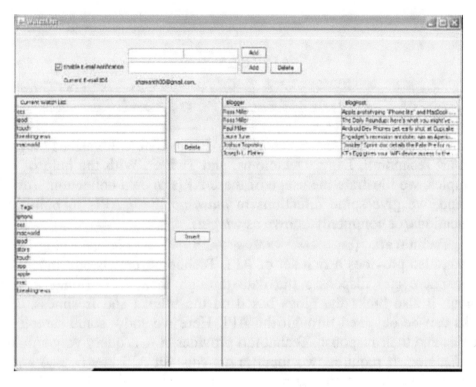

Figure A.6: Watchlist feature of the BlogTrackers that helps in monitoring future occurrences of "hot" keywords in blog posts.

API Examples

Here we present a few examples of the APIs available at various blog sites such as Technorati, Digg, Del.icio.us, and Twitter. With the help of these examples, we illustrate the usage of these APIs in data collection. Towards the end, we give some directions to *mash-up* these APIs to build small applications, or commonly known as *widgets*.

Technorati, (http://technorati.com/developers/api/) a popular blog search engine, also provides a rich set of API. Technorati crawls and indexes the blogosphere and dispenses the data through their API. Being a search engine, it also ranks the blogs based on the inlinks and freshness. These ranks can be obtained through the API. Here we show some sample API queries and their response. Technorati provides bloginfo query very similar to BlogCatalog. It requires two input arguments: an API key (apikey) and the blog **url.** This request would return the name of the blog, blog URL, blog RSS URL, inbound blogs, inbound links, rank, language, and foaf URL. Foaf stands for "friends-of-a-friend" and is the URL for the XML format file that specifies the social network information of the blogger.

Example B.1. The sample request for bloginfo looks like:

```
http://api.technorati.com/bloginfo?key=[apikey]&url= http://
blog.johndoe.com
```

and the sample response looks like:

```
<result>
 <url>http://blog.johndoe.com</url>
 <weblog>
   <name>The John Doe Blog</name>
   <url>http://blog.johndoe.com</url>
   <rssurl>http://blog.johndoe.com/rss/</rssurl>
   <atomurl>[blog Atom URL]</atomurl>
   <inboundblogs>23</inboundblogs>
   <inboundlinks>40</inboundlinks>
   <lastupdate>2009-04-30</lastupdate>
   <rank>738</rank>
   <lang>English</lang>
   <foafurl>http://blog.johndoe.com/foaf/</foafurl>
 </weblog>
</result>
```

Another query provided by Technorati is the blogposttags query. It requires two input arguments: an API key (apikey) and the blog url. The query can also take a third optional parameter, i.e., the limit parameter (querycount) that governs how many tags should be returned. This request would return a list of tag names and a count with each tag which denotes the number of times the tag appears in the blog.

Example B.2. The sample request for blogposttags looks like:
```
http://api.technorati.com/blogposttags?key=[apikey]&url= http://
blog.johndoe.com&querycount=1
```

and the sample response looks like:
```
<result>
 <querycount>1</querycount>
 <item>
   <tag>camera</tag>
   <posts>10</posts>
 </item>
</result>
```

Digg,(http://apidoc.digg.com/) another popular social media website, is all about user rating. Users rate the stories collaboratively. More popular stories are ranked higher. The user community helps in deciding which stories are worth reading. Digg provides a rich set of API that can be used to retrieve the popular stories appearing at a blog for a particular time period. Here we illustrate Digg API through an example query and the sample response. Digg provides a "list stories" API that can be used to retrieve popular or most recent blog posts for a specified time period. We illustrate the popular story query with the sample response. It requires 5 input arguments: the blog URL (domain), number of stories the reader is interested in (count), time period start date (min-submit-date), time period end date (max-submit-date), and API key (appkey). This request would return a list top-k (k is specified by the user through the argument count) popular stories from the blog URL specified by the domain argument. The time period of the stories is governed by the min-submit-date and max-submit-date.

Example B.3. The sample query for popular story looks like:
```
http://services.digg.com/stories/popular?domain= blog.johndoe.com &
count=10 &min-submit-date=epoch(07/01/2008) &max-submit-date=
epoch(07/15/1008) &appkey=[apikey]
```

Note that the start and the end time periods are specified in epoch. This sample query lists 10 most popular stories from Engadget (http://www.engadget.com/) between July 1st, 2008 and July 15th, 2008. The sample response looks like:

```
<story id="7511382" link="http://www.engadget.com/2008/07/15/dev-
 team-shows-off-video-of-worlds-first-jailbroken-iphone-3g/"
 submit-date="1216139955" diggs="623" comments="38"
 promote-date="1216186807" status="popular" media="news"
  href="http://digg.com/apple/World-s-First-Jailbroken-iPhone-3G">
   <title>World's First Jailbroken iPhone 3G</title>
   <description>
    We can't say this is a surprise. . . but it is sweet to see.
    The iPhone Dev Team has added a video to their blog
    showing off the latest version of their upcoming PwnageTool
    2.0, along with a video of what they claim is the "world's
first"
    jailbroken iPhone 3G.
   </description>
   <user name="jordankasteler" icon="http://digg.com/users/
    jordankasteler/l.png" registered="1172914233" profileviews=
    "8344" fullname="JordanKasteler"/>
   <topic name="Apple" short-name="apple"/>
   <container name="Technology" short-name="technology"/>
   <thumbnail originalwidth="500" originalheight="378" content
   Type=
   "image/jpeg" src="http://digg.com/apple/World-s-First-
   Jailbroken-iPhone-3G/t.jpg" width="80" height="80"/>
</story>
 .
 .
 .
```

Delicious, (http://delicious.com/help/api) a social bookmarking service that allows users to tag, save, manage and share web pages from a centralized source. Delicious helps people discover, remember and share on the Internet with the emphasis on the power of the community. Delicious APIs provide access to the bookmarks (website, blogs, etc.) and the tags contributed by the community. Delicious APIs include functions for post, tags, and tag bundles. API queries under post can be used to obtain the bookmarked posts by a specific user. Queries under tags can be used to obtain the tags contributed by a specific user. Queries under tag bundles can be used to obtain all the tags contributed by a user to a bundle of media such as websites, blogs, etc. A bundle of media refers to a category to which all the websites, blogs, etc. are assigned. For example, a "music" bundle could contain all the websites, blogs, etc. that are bookmarked under music category by the user. Here we illustrate a query from tags API, get through an example and sample response.

Example B.4. The sample query for get looks like:
```
https://api.del.icio.us/v1/tags/get
```

where v1 is a user. The query returns a list of tags and number of times used by a user, v1. Note that Delicious APIs do not require an API key; however, it is required to have a one second gap between two consecutive queries to prevent throttling (a term commonly refers to blocking the requests by the server). The sample response of the above query looks:

```
<tags>
 <tag count="1" tag="activedesktop" />
 <tag count="1" tag="business" />
 <tag count="3" tag="radio" />
 <tag count="5" tag="xml" />
 <tag count="1" tag="xp" />
 <tag count="1" tag="xpi" />
</tags>
```

Twitter, (http://apiwiki.twitter.com/) a microblogging social media service for friends, family, and co-workers to communicate and stay connected through the exchange of quick, frequent answers to one simple question: "What are you doing?". Time magazine mentioned that "Twitter is on its way to become the next killer app". New York Times mentioned that "It's one of the fastest-growing phenomena on the Internet". It allows users to answer that question in 140 characters. This constraint forces the users to be concise and informative. The answers to the question are considered as status updates of the members, which are timestamped. A member can be followed by other members who can watch his(her) tweets. These members are called "followers". A member can also follow other members and see their updates or "tweets". Those members whom you follow are considered your "friends". This constructs a directional friendship network, i.e., the members whom I follow are my friends, but they may not necessarily follow me so I am not in their friend network.

Twitter provides an extremely rich set of APIs to search the status messages, or the tweets and various timeline, status, user, friendship, and social graph methods to obtain the tweets, profile details, friends, and followers of a specific member. Here we present sample API queries and their response. We start with the statuses/show method. It shows a single status message provided by a member. It also shows the member details in the response. It requires one input argument, the id of the status you wish to retrieve.

Example B.5. The sample query for statuses/show looks like:

```
http://twitter.com/statuses/show/1472669360.xml
```

Note that twitter API does not require an API key so it restricts the access to 1 call per request. Also, only members who have posted their tweets publicly (or commonly known as non-protected) are available for download. The sample response generated by the above query looks like:

```
<?xml version="1.0" encoding="UTF-8"?>
<status>
 <created-at>Tue Apr 07 22:52:51 +0000 2009</created-at>
 <id>1472669360</id>
 <text>
   At least I can get your humor through tweets. RT abdur:
   I don't mean this in a bad way, but genetically speaking
   your a
   cul-de-sac.
 </text>
 <user>
   <id>1401881</id>
   <name>Doug Williams</name>
   .
   .
   .
   <protected>false</protected>
   <followers-count>1027</followers-count>
   <friends-count>293</friends-count>
   <statuses-count>3390</statuses-count>
   .
   .
   .
 </user>
 </status>
```

statuses/user_timeline can be used to download 20 most recent status messages of a member. Next, we illustrate the friends/ids query of the twitter API. It takes user_id or screen_name of the user and returns the list of friends or the members (s)he is following.

Example B.6. The sample query for friends/ids looks like:

```
http://twitter.com/friends/ids/bob.xml
```

and the sample response looks like:

```
<?xml version="1.0" encoding="UTF-8"?>
<ids>
 <id>1401881</id>.
 <id>6761692</id>
 <id>6636732</id>
 <id>813286</id>
 <id>7057722</id>
 .
 .
 .
</ids>
```

Similarly, the followers/ids query gives the followers of a user specified by user_id or screen_name.

Example B.7. The sample query for followers/ids looks like:

```
http://twitter.com/followers/ids/bob.xml
```

and the sample response looks like:

```
<?xml version="1.0" encoding="UTF-8"?>
<ids>
 <id>683643</id>
 <id>744883</id>
 <id>755002</id>
 <id>611823</id>
 .
 .
 .
</ids>
```

It is rather more interesting when one can build applications that can interact with different data sources at different sites. With the phenomenal growth of social media services also comes a growing list of site-specific APIs that developers must learn to build applications and mashups that talk to different data sources. Google's **OpenSocial** (http://code.google.com/apis/opensocial/) is an effort in this direction. It defines a common API for social applications or mashups across multiple websites. It relies on standard JavaScript and HTML constructs using which developers can create apps that access a social network's friends and update feeds. Yahoo **Pipes** (http://pipes.yahoo.com/pipes/) is a similar composition tool that aggregates and manipulates content from multiple websites. Developers can specify the URL, of the data source including the API query URL and yahoo pipes will grab the output as RSS, JSON, and other compatible formats, and combine them into one. It also provides sorting, filtering, and translating mechanisms.

Bibliography

[1] T. O'Reilly. What is Web 2.0 - design patterns and business models for the next generation of software. http://www.oreillynet.com/pub/a/oreilly/tim/news/2005/09/30/what-isweb-20.html, September 2005.

[2] D. Gillmor. *We the Media: Grassroots Journalism by the People, for the People*. O'Reilly, 2006.

[3] R. Scoble and S. Israel. *Naked conversations: how blogs are changing the way businesses talk with customers*. John Wiley, 2006.

[4] M. Richardson and P. Domingos. Mining knowledge-sharing sites for viral marketing. In *Proceedings of the Eighth ACM SIGKDD International Conference on Knowledge Discovery and Data Mining*, pages 61–70, New York, NY, USA, 2002. ACM Press. DOI: 10.1145/775047.775057

[5] D. Gruhl, R. Guha, R. Kumar, J. Novak, and A. Tomkins. The predictive power of online chatter. In *Proceeding of the Eleventh ACM SIGKDD International Conference on Knowledge Discovery in Data Mining*, pages 78–87, New York, NY, USA, 2005. ACM Press. DOI: 10.1145/1081870.1081883

[6] G. Mishne and M. Rijke. Deriving wishlists from blogs show us your blog, and we'll tell you what books to buy. In *Proceedings of the Fifteenth International Conference on World Wide Web*, pages 925–926, New York, NY, USA, 2006. ACM Press. DOI: 10.1145/1135777.1135947

[7] T. Coffman and S. Marcus. Dynamic Classification of Groups Through Social Network Analysis and HMMs. In *Proceedings of IEEE Aerospace Conference*, pages 3197 – 3205, 2004.

[8] Mike Thelwall. Bloggers under the London attacks: Top information sources and topics. In *Proceedings of the Third Annual Workshop on Webloging Ecosystem: Aggreation, Analysis and Dynamics*, 2006.

[9] N. Agarwal, H. Liu, S. Murthy, A. Sen, and X. Wang. A social identity approach to identify familiar strangers in a social network. In *Proceedings of the Third International AAAI Conference of Weblogs and Social Media*, pages 2–9, 2009.

[10] J. Q. Fang and Y Liang. Topological properties and transition features generated by a new hybrid preferential model. *Chin. Phys. Lett.*, 22(10):2719–2722, 2005. DOI: 10.1088/0256-307X/22/10/072

[11] F. Chung and L. Lu. Complex graphs and networks. In *Cbms Regional Conference Series in Mathematics*, 2006.

[12] Vicenç Gómez, Andreas Kaltenbrunner, and Vicente López. Statistical analysis of the social network and discussion threads in slashdot. In *Proceeding of the 17th international conference on World Wide Web (WWW)*, pages 645–654, 2008. DOI: 10.1145/1367497.1367585

[13] A. Ng, M. Jordan, and Y Weiss. *On spectral clustering: analysis and an algorithm*. Advances in Neural Information Processing Systems 14. MIT Press. (Ed: T. Dietterich and S. Becker and Z. Ghahramani), 2002.

[14] J. Shi and J. Malik. Normalized cuts and image segmentation. *IEEE Transactions on Pattern Analysis and Machine Intelligence*, 22(8):888–905, 2000. DOI: 10.1109/34.868688

[15] L. Hagen and A.B. Kahng. New spectral methods for ratio cut partitioning and clustering. *IEEE Transactions on Computer-Aided Design of Integrated Circuits and Systems*, 11(9):1074–1085, 1992. DOI: 10.1109/43.159993

[16] U. Luxburg. A tutorial on spectral clustering. *Statistics and Computing*, 17(4):395–416, 2007. DOI: 10.1007/s11222-007-9033-z

[17] Y. Chi, X. Song, D. Zhou, K. Hino, and B.L. Tseng. Evolutionary spectral clustering by incorporating temporal smoothness. In *Proceedings of the 13th ACM SIGKDD international conference on*

Knowledge discovery and data mining, pages 153–162. ACM New York, NY, USA, 2007. DOI: 10.1145/1281192.1281212

[18] H. Ning, W. Xu, Y. Chi, Y. Gong, and T. Huang. Incremental spectral clustering with application to monitoring of evolving blog communities. In *SIAM International Conference on Data Mining*, 2007.

[19] Y. Chi, S. Zhu, X. Song, J. Tatemura, and B.L. Tseng. Structural and temporal analysis of the blogosphere through community factorization. In *Proceedings of the 13th ACM SIGKDD international conference on Knowledge discovery and data mining*, pages 163–172. ACM New York, NY, USA, 2007. DOI: 10.1145/1281192.1281213

[20] B. Li, S. Xu, and J. Zhang. Enhancing clustering blog documents by utilizing author/reader comments. In *Proceedings of the 45th Annual Southeast Regional Conference*, pages 94–99, New York, NY, USA, 2007. ACM Press. DOI: 10.1145/1233341.1233359

[21] C.H. Brooks and N. Montanez. Improved annotation of the blogosphere via autotagging and hierarchical clustering. In *Proceedings of the Fifteenth International Conference on World Wide Web*, pages 625–632, New York, NY, USA, 2006. ACM Press. DOI: 10.1145/1135777.1135869

[22] N. Agarwal, M. Galan, H. Liu, and S. Subramanya. Clustering blogs with collective wisdom. In *Proceedings of the International Conference on Web Engineering*, 2008. DOI: 10.1109/ICWE.2008.9

[23] A. Java, A. Joshi, and T. Finin. Detecting commmunities via simultaneous clustering of graphs and folksonomies. In *Proceedings of the Tenth Workshop on Web Mining and Web Usage Analysis (WebKDD)*. ACM, 2008.

[24] Q. Xu, M. desJardins, and K. Wagstaff. Active constrained clustering by examining spectral eigenvectors. *Discovery Science*, pages 294–307, 2005. DOI: 10.1007/11563983_25

[25] E. Keller and J. Berry. *One American in ten tells the other nine how to vote, where to eat and, what to buy. They are The Influentials*. The Free Press, 2003.

[26] T. Elkin. Just an online minute . . . online forecast. http://publications.mediapost.com/index.cfm?fuseaction =Articles.showArticle art aid=29803.

[27] D. Drezner and H. Farrell. The power and politics of blogs. In *American Political Science Association Annual Conference*, 2004. DOI: 10.1007/s11127-007-9198-1

[28] S. Brin and L. Page. The anatomy of a large-scale hypertextual Web search engine. In *Proceedings of the Seventh International Conference on World Wide Web*, pages 107 – 117, 1998. DOI: 10.1016/S0169-7552(98)00110-X

[29] J. Kleinberg. Authoritative sources in a hyperlinked environment. In *9th ACM-SIAM Symposium on Discrete Algorithms*, 1998. DOI: 10.1145/324133.324140

[30] A. Kritikopoulos, M. Sideri, and I. Varlamis. Blogrank: ranking weblogs based on connectivity and similarity features. In *Proceedings of the Second International Workshop on Advanced Architectures and Algorithms for Internet Delivery and Applications*, page 8, New York, NY, USA, 2006. ACM Press. DOI: 10.1145/1190183.1190193

[31] K.E. Gill. How can we measure the influence of the blogosphere? In *Proceedings of the Workshop on the Weblogging Ecosystem: Aggregation, Analysis and Dynamics*, 2004.

[32] M. Faloutsos, P. Faloutsos, and C. Faloutsos. On power-law relationships of the internet topology. In *Proceedings of the conference on Applications, technologies, architectures, and protocols for computer communication*, pages 251–262, 1999. DOI: 10.1145/316188.316229

[33] C. Anderson. *The long tail : why the future of business is selling less of more*. New York: Hyperion, 2006.

[34] J. Leskovec, M. McGlohon, C. Faloutsos, N. Glance, and M. Hurst. Cascading behavior in large blog graphs. In *SIAM International Conference on Data Mining*, 2007.

[35] J. Goldenberg, B. Libai, and E. Muller. Talk of the network: A complex systems look at the underlying process of word-of-mouth. *Marketing Letters*, 12:211 – 223, 2001. DOI: 10.1023/A:1011122126881

[36] D. Gruhl, David Liben-Nowell, R. Guha, and A. Tomkins. Information diffusion through blogspace. *SIGKDD Exploration Newsletter*, 6(2):43–52, 2004. DOI: 10.1145/1046456.1046462

[37] D. Kempe, J. Kleinberg, and E. Tardos. Maximizing the spread of influence through a social network. In *Proceedings of the International Conference on Knowledge Discovery and Data Mining*, pages 137–146, New York, NY, USA, 2003. ACM Press. DOI: 10.1145/956750.956769

[38] A. Java, P. Kolari, T. Finin, and T. Oates. Modeling the spread of influence on the blogosphere. In *Proceedings of the Fifteenth International World Wide Web Conference*, 2006.

[39] Jure Leskovec, Andreas Krause, Carlos Guestrin, Christos Faloutsos, Jeanne VanBriesen, and Natalie Glance. Cost-effective outbreak detection in networks. In *Proceedings of the 13th ACM SIGKDD international conference on Knowledge discovery and data mining*, pages 420–429, New York, NY, USA, 2007. ACM. DOI: 10.1145/1281192.1281239

[40] S. Nakajima, J. Tatemura, Y. Hino, Y. Hara, and K. Tanaka. Discovering important bloggers based on analyzing blog threads. In *Annual Workshop on the Weblogging Ecosystem*, 2005.

[41] D.J. Watts and P.S. Dodds. Influentials, networks, and public opinion formation. *JOURNAL OF CONSUMER RESEARCH*, 34(4):441, 2007. DOI: 10.1086/518527

[42] D.J. Watts. Challenging the influentials hypothesis. *WOMMA Measuring Word of Mouth*, 3:201–211, 2007.

[43] J. Zhuang, S.C.H. Hoi, A. Sun, and R. Jin. Representative entry selection for profiling blogs. In *Proceedings of the International Conference on Knowledge Management (CIKM)*. ACM New York, NY, USA, 2008. DOI: 10.1145/1458082.1458293

[44] N. Agarwal, H. Liu, L. Tang, and P.S. Yu. Identifying the influential bloggers in a community. In *Proccedings of the First ACM International Conference on Web Search and Data Mining (Video available at: http://videolectures.net/wsdm08_agarwal_iib/)*, pages 207 – 218, 2008. DOI: 10.1145/1341531.1341559

[45] G.D. Fensterer. *Planning and Assessing Stability Operations: A Proposed Value Focus Thinking Approach*. PhD thesis, Air Force Institute of Technology, 2007.

[46] R. L. Keeney and H. Raiffa. *Decisions with Multiple Objectives: Preferences and Value Tradeoffs*. Cambridge University Press, 1993.

[47] R. Motwani and P. Raghavan. *Randomized Algorithms*. Cambridge University Press, 1995.

[48] G.H. Golub and C.F. Van Loan. *Matrix Computations*. Johns Hopkins University Press, 3rd edition edition, 1996.

[49] X. Ni, G. Xue, X. Ling, Y. Yu, and Q. Yang. Exploring in the weblog space by detecting informative and affective articles. In *Proceedings of the 16th International Conference on World Wide Web*, pages 281–290, New York, NY, USA, 2007. ACM. DOI: 10.1145/1242572.1242611

[50] X. Song, Y. Chi, K. Hino, and B. Tseng. Identifying opinion leaders in the blogosphere. In *Proceedings of the Sixteenth ACM International Conference on Information and Knowledge Management*, pages 971–974, New York, NY, USA, 2007. ACM. DOI: 10.1145/1321440.1321588

[51] P. Sztompka. *Trust: A sociological theory*. Cambridge Univ Pr, 1999.

[52] R. Guha, R. Kumar, P. Raghavan, and A. Tomkins. Propagation of trust and distrust. In *Proceedings of the Thirteenth International Conference on World Wide Web*, pages 403–412, New York, NY, USA, 2004. ACM Press. DOI: 10.1145/988672.988727

[53] L. Terveen and D.W. McDonald. Social matching: A framework and research agenda. *ACM Transactions on Computer Human Interaction*, 12(3):401–434, 2005. DOI: 10.1145/1096737.1096740

[54] J. Golbeck and J. Hendler. Inferring binary trust relationships in web-based social networks. *ACM Transactions on Internet Technology*, 6(4):497–529, 2006. DOI: 10.1145/1183463.1183470

[55] A. Kale, A. Karandikar, P. Kolari, A. Java, T. Finin, and A. Joshi. Modeling trust and influence in the blogosphere using link polarity. In *International Conference on Weblogs and Social Media*, 2007.

[56] P. Dwyer. Building trust with corporate blogs. In *International Conference on Weblogs and Social Media*, 2007.

[57] N. Agarwal and H. Liu. Trust in blogosphere. Encyclopedia of Database Systems, August 2009.

[58] Y. Lin, H. Sundaram, Y. Chi, J. Tatemura, and B. Tseng. Detecting splogs via temporal dynamics using self-similarity analysis. *ACM Transactions of the Web*, 2(1), 2008. DOI: 10.1145/1326561.1326565

[59] Z. Gyongyi, H. Garcia-Molina, and J. Pedersen. Combating web spam with trustrank. In *Proceedings of the 30th International Conference on Very Large Data Bases (VLDB)*, 2004.

[60] L. Zhu, A. Sun, and B. Choi. Online spam-blog detection through blog search. In *Proceedings of the Seventeenth ACM International Conference on Information and Knowledge Management (CIKM)*, pages 1347–1348, 2008. DOI: 10.1145/1458082.1458272

[61] J. Leskovec, J. Kleinberg, and C. Faloutsos. Graphs over time: densification laws, shrinking diameters and possible explanations. In *Proceeding of the ACM SIGKDD international conference on Knowledge Discovery in Data mining (KDD)*, pages 177–187, 2005. DOI: 10.1145/1081870.1081893

[62] E. Kamaliha, F. Riahi, V. Qazvinian, and J. Adibi. Characterizing network motifs to identify spam comments. In *IEEE International Conference on Data Mining Workshops, 2008. ICDMW'08*, pages 919–928, 2008. DOI: 10.1109/ICDMW.2008.72

[63] A. Ntoulas, M. Najork, M. Manasse, and D. Fetterly. Detecting spam web pages through content analysis. In *Proceedings of the 15th*

international conference on World Wide Web, pages 83–92, 2006. DOI: 10.1145/1135777.1135794

[64] G. Mishne, D. Carmel, and R. Lempel. Blocking blog spam with language model disagreement. In *Proceedings of the First International Workshop on Adversarial Information Retrieval on the Web (AIRWeb)*, 2005.

[65] Z. Gyongyi, P. Berkhin, Hector Garcia-Molina, and J. Pedersen. Link spam detection based on mass estimation. In *Proceedings of the 32nd International Conference on Very Large Data Bases (VLDB)*, 2006.

[66] P. Kolari, A. Java, T. Finin, T. Oates, and A. Joshi. Detecting spam blogs: A machine learning approach. In *Proceedings of the 21st National Conference on Artificial Intelligence (AAAI)*, 2006.

[67] K. Burton, A. Java, and I. Soboroff. The icwsm 2009 spinn3r dataset. In *Proceedings of the Third Annual Conference on Weblogs and Social Media (ICWSM)*, 2009.

[68] B. Salton and C. Buckley. Term weighting approaches in automatic text retrieval. *Information Processing and Management*, 24(5):513–523, 1988. DOI: 10.1016/0306-4573(88)90021-0

[69] H. Liu and H. Motoda. *Feature selection for knowledge discovery and data mining*. Springer, 1998.

[70] S. Deerwester, S.T. Dumais, G.W. Furnas, T.K. Landauer, and R. Harshman. Indexing by latent semantic analysis. *Journal of the American Society for Information Science*, 41:391–407, 1990. DOI: 10.1002/(SICI)1097-4571(199009)41:6<391::AID-ASI1>3.0.CO;2-9

[71] T. Hofmann. Probabilistic latent semantic indexing. In *Proceedings of the 22nd annual international ACM SIGIR conference on Research and development in information retrieval*, pages 50–57. ACM New York, NY, USA, 1999. DOI: 10.1145/312624.312649

[72] D.M. Blei, A.Y. Ng, and M.I. Jordan. Latent dirichlet allocation. *The Journal of Machine Learning Research*, 3:993–1022, 2003. DOI: 10.1162/jmlr.2003.3.4-5.993

[73] S. Prevost. An information structural approach to spoken language generation. In *Proceedings of the 34th annual meeting on Association for Computational Linguistics*, pages 294–301. Association for Computational Linguistics Morristown, NJ, USA, 1996. DOI: 10.3115/981863.981902

[74] C. Hayes and P. Avesani. Using tags and clustering to identify topic-relevant blogs. In *International Conference on Weblogs and Social Media*, 2007.

[75] N. Agarwal, S. Kumar, H. Liu, and M. Woodward. Blogtrackers: A tool for sociologists to track and analyze blogosphere. In *Proceedings of the Third International AAAI Conference on Weblogs and Social Media (ICWSM)*, pages 359 – 360, 2009.

[76] N. Bansal and N. Koudas. Searching the blogosphere. In *10th International Workshop on the Web and Databases (WebDB 2007)*, 2007.

[77] J Lee. Exploring global terrorism data. *ACM Crossroads*, 15(2):7–14, 2008. DOI: 10.1145/1519390.1519393

Biography

NITIN AGARWAL

Nitin Agarwal is a professor of Information Science at University of Arkansas at Little Rock. He received his Bachelor of Technology in Information Technology from Indian Institute of Information Technology, India, and Ph.D. in Computer Science from Arizona State University. He is one of the founding members of the Social Computing group in the Data Mining and Machine Learning Lab at ASU. His primary research interests include Social Computing, Knowledge Extraction in Social Media, Modeling Influence, Collective Wisdom, Familiar Strangers, and Model Evaluation. His work has resulted in publications in various prestigious forums including book chapters, encyclopedia entries, conferences and journals. His presentation at Web Search and Data Mining (WSDM 2008) conference on "Identifying the Influential Bloggers in a Community" recorded the highest number of hits (over 700) among all the talks at the conference (http://videolectures.net/wsdm08_agarwal_iib/). He co-presented a tutorial at the premiere data mining conference KDD 2008 on "Blogosphere: Research Issues, Applications, and Tools" (http://videolectures.net/kdd08_liu_briat/). He is a co-guest editor of a special issue on "Social Computing in Blogosphere" for IEEE Internet Computing magazine appearing (2010).

HUAN LIU

Huan Liu is a professor of Computer Science and Engineering at Arizona State University. He received his Bachelor of Engineering from Shanghai Jiao Tong University and Ph.D. from University of Southern California, researched at Telecom Research Labs in Australia, and taught at National University of Singapore before he joined ASU. He has been recognized for excellence in teaching and research in CSE, ASU. His research interests are in data/web mining, machine learning, social computing, and artificial intelligence, investigating problems that arise in many real-world applications with high-dimensional data of disparate forms such as social media, modeling group interaction, text categorization, biomarker identification, and text/web mining. His research is sponsored by NSF, NASA, AFOSR, and ONR, among others. His well-cited publications include books, book chapters, encyclopedia entries as well as conference and journal papers. He serves on journal editorial boards and numerous conference program committees, and is a founding organizer of the International Workshop Series on Social Computing, Behavioral Modeling, and Prediction (http://sbp.asu.edu/) in Phoenix, AZ (SBP'08 and SBP'09). His professional memberships include AAAI, ACM, ASEE, and IEEE.

Index

The index that appeared in the print version of this title was intentionally removed from the eBook. Please use the search function on your eReading device for terms of interest. For your reference, the terms that appear in the print index are listed below

Printed in the United States
by Baker & Taylor Publisher Services